Fertility of the Sea

Fertility of the Sea

Volume 2

Edited by
JOHN D. COSTLOW, Jr.
Duke University, Marine Laboratory
Beaufort, North Carolina

GORDON AND BREACH SCIENCE PUBLISHERS

New York   London   Paris

551.46
Sy6t
85138
oct 1973

Foreword

The Symposium on Fertility of the Sea, the ninth in a continuing series of symposia sponsored by Latin American universities, was held in Sao Paulo, Brazil, December 1–6, 1969, under the Chairmanship of Dra. Marta Vannucci of the Oceanographic Institute of the University of Sao Paulo. This was an important symposium of great scientific and social significance and brought together many scientists from all over the world, some from as far as India. This was made possible through the generous support of international agencies such as UNESCO, FAO, The Royal Society, and foreign governments. Excellent cooperation from the U. S. Atomic Energy Commission and the National Science Foundation helped to ensure the success of the meeting. The generous support of the Ford Foundation through a grant to the National Academy of Sciences is gratefully acknowledged. As in previous symposia, the Oak Ridge National Laboratory was a co-sponsor. The Marine Laboratory of Duke University was the U. S. university co-sponsor.

A symposium on oceanography was originally suggested by Dr. K. N. Rao of the Ford Foundation, and was developed through discussions with Dra. Vannucci and with Drs. Karl Wilbur and John Costlow of Duke University.

The program was arranged to present a picture of the current status of studies on fertility of the sea, a subject of great economic and social importance. The excellent cooperation of Brazilian authorities, from both the National Research Council and the University of Sao Paulo, and the invitation to hold the meeting in the beautiful new Architectural School, left a very good impression on all the visitors and participants.

A sufficient number of important questions were brought up during the course of the meeting to necessitate additional sessions on Saturday afternoon after the official close of the symposium. During these Saturday afternoon sessions a workshop was planned for the following week. Emphasized during this workshop was the need for the development of a number of tropical oceanography stations in the Southern Americas, and the concomitant need of centers for training young people in tropical oceanography to staff these future stations.

The writer, while not in the field of oceanography, was most pleased with the excellent cooperation and the international character of this meeting. The published proceedings should form a milestone in the area of oceanography.

Again, we want to thank out Latin American colleagues for sponsoring the symposium and for making it a most remarkable meeting.

Further symposia in this series are as follows:

1961

International Symposium on Tissue Transplantation, Santiago, Viña del Mar, and Valparaiso, Chile (published in 1962 by the University of Chile Press, Santiago; Edited by A. P. Cristoffanini and Gustavo Hoecker; 269 pp.).

1962

Symposium on Mammalian Cytogenetics and Related Problems in Radiobiology, Sao Paulo and Rio de Janeiro, Brazil (published in 1964 by The MacMillan Company, New York, under arrangement with Pergamon Press, Ltd., Oxford; Edited by C. Pavan, C. Chagas, O. Frota-Pessoa, and L. R. Caldas; 427 pp.).

1963

International Symposium on Control of Cell Division and the Induction of Cancer, Lima, Peru, and Cali, Columbia (published in 1964 by the U. S. Department of Health, Education, and Welfare as National Cancer Institute Monograph 14; Edited by C. C. Congdon and Pablo Mori-Chavez; 403 pp.).

1964

International Symposium on Genes and Chromosomes, Structure and Function, Buenos Aires, Argentina (published in 1965 by the U. S. Department of Health, Education, and Welfare as National Cancer Institute Monograph 18; Edited by J. I. Valencia and Rhoda F. Grell, with the cooperation of Ruby Marie Valencia; 354 pp.).

1965

International Symposium on the Nucleolus—Its Structure and Function, Montevideo, Uruguay (published in 1966 by the U. S. Department of Health, Education, and Welfare as National Cancer Institute Monograph 23; Edited by W. S. Vincent and O. L. Miller, Jr.; 610 pp.).

1966

International Symposium on Enzymatic Aspects of Metabolic Regulation, Mexico City, Mexico (published in 1967 by the U. S. Department of Health, Education, and Welfare as National Cancer Institute Monograph 27; Edited by M. P. Stulberg; 343 pp.).

1967

International Symposium on Basic Mechanisms in Photochemistry and Photobiology, Caracas, Venezuela (published in 1968 by Pergamon Press as Volume 7, Number 6, *Photochemistry and Photobiology;* Edited by James W. Longworth; 326 pp.).

1968

International Symposium on Nuclear Physiology and Differentiation, Belo Horizonte, Minas Gerais, Brazil (published in 1969 by the Genetics Society of America as a supplement to *Genetics,* Volume 61, No. 1, Edited by Robert P. Wagner; 469 pp.).

ALEXANDER HOLLAENDER

Biology Division,
Oak Ridge National Laboratory*
Oak Ridge, Tennessee, U.S.A.

---

* Operated by Union Carbide Corporation for the U. S. Atomic Energy Commission.

Contents

Contents

The relationship between chaetognaths, water masses, and standing stock off the Colombia Pacific coast

FRANCISCO PINEDA POLO

Universidad del Valle
Cali, Colombia

Abstract

The present report tries to show the relationship between chaetognath distribution and surface water masses in the zone Buenaventura Isla Gorgona and Tumaco (Colombia). The taxonomic study of the samples has not been *achieved* but analysis of temperature and salinity compared to chaetognath distribution indicates that in this region there are two planktonic communities:
1. a diluted shallow water community and
2. an oceanic or equatorial-pacific community.

These two communities are marked by different species of chaetognaths. It has also been possible, in a preliminary way, to determine the relationship between the two water masses mentioned above and the relative abundance of zooplankton.

Resumen

O presente trabalho, tenta mastrar as interrelações existentes entre a distribuição dos quetognatos e massas de água de superfície, na zona de Buenaventura-Ilha Gorgona e Tumaco (Colombia). O estudo taxonômico das amostras não foi completado mas as análises de temperatura e salinidade comparadas com a distribuição de quetognatos indica que nessas regiões duas comunidades planctônicas ocorrem:
1. Uma comunidade de águas rasas, diluida.
2. Uma comunidade Oceânica ou Pacífico-Equatorial.

Essas duas comunidades são caracterizadas por espécies diferentes de quetognatos.

Foi possível determinar as interrelações entre abundância de zooplâncton e as duas massas de água.

MATERIALS AND METHODS

Since January 1969 I have been taking plankton samples every month from the described area. At the same time temperature, salinity, and oxygen determinations have been made. However, only the results for April, May, July and August are presented because we consider these are the most representative.

The plankton hauls were taken with a Norpac plankton net of 0.45 mts diameter at the mouth, 1.80 mts long and 0.1061 mm mesh size. A plastic cod-end has been used. I have chosen a definite time unit of 15 seconds for each haul, in order to offset the lack of a flow-meter. The cable length is 30 m approximately. The net has been towed from a manual winch. Each sample was labeled and preserved in 5% formalin.

Temperature was measured *in situ* with reversing thermometers coupled to Nansen bottles. The salinity was measured by the Knudsen–Morh method and the oxygen by the Winkler method.

All the chaetognaths have been removed from the samples and examined with stereomicroscope and, on occasion, with a light microscope.

For the zooplankton countings I divided each sample in 140 cc. aliquots, taking $1/4$ of one aliquot with a Cushing Splitter. For countings we used a stereomicroscope and a Planktin Sorting Tray.

DESCRIPTION OF THE SURFACE TEMPERATURE-SALINITY AND OXYGEN VARIATIONS

The waters of this region could be separated into two water masses of different origin: one comes from the equatorial pacific water, with high salinity ($35^0/_{00}$) and a temperature range of 27–28°C. The second one is the water above the continental shelf with important salinity fluctuations caused by the rivers run-off, and temperature which in certain local areas could rise to 29°C. At the mouth of the rivers salinity is particularly low and temperatures higher.

Considering that this is the first work on the littoral zone of the Colombian-Pacific Coast, there is no literature available to compare with my results. The Atlas of Temperature and Salinity published by the U.S.N. Oceanographic Office give a general idea of the isotherms and isohalines for the region. Bennett (1966) gives only a general draft of the isohalines in the Panama Bight.

April 4–7, 1969, Cruise C

The temperatures for this month were measured for the sector between Buenaventura and Punta Coco, by six stations only, and do not allow any speculation about distribution. However, we observe a concentration of high temperatures (Fig. 1, Fig. 2) near the shore.

Salinity for this cruise, (Fig. 3) shows the same difficulty that was mentioned above in the case of temperature. We notice, however, that the zone near the coast line has low salinity water mainly at the mouth of the rivers Raposso, Cajambre, Naya and Guapi. These rivers contribute permanently the nutrients responsible for the relative abundance of the plankton in the nearshore zone. The isohaline of $35^o/_{oo}$ is near the coast and may be due to the low rain index during this period and consequently a low river run-off.

May 16–17, 1969, Cruise D

Eight stations we have made during this month. They have been situated between Buenaventura, Punta Coco and Gorgona Island (Fig. 1).

The temperature shows approximately the same distribution as during April. The highest temperature values are found near the shore line and the lowest temperature off shore (Fig. 4).

The salinity (Fig. 5) goes down progressively approaching the coast. The stations D-6 and D-7, with salinities of $31.1^o/_{oo}$ and $32.9^o/_{oo}$ respectively (Table 1), mark a water of "transition type", which has been confirmed by the analysis of plankton found in it. This progression of the coastal waters to the west may be the result of the high precipitation during this month associated with the increase in rivers run-off. ·

July 1–3, 1969, Cruise F

During this month the temperature shows a similar distribution: i.e. high temperatures near the coast line (28°C approximately) (Fig. 6) and low temperatures off shore. The isotherms tendency to approach the coast may

Fertility of the Sea

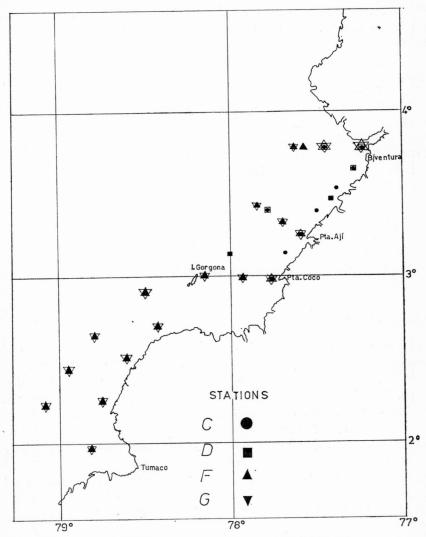

Figure 1 Localization of the stations for Cruise C, D, F, and G, between
Buenaventura, Gorgona Island and Tumaco (Colombia, S.A.)

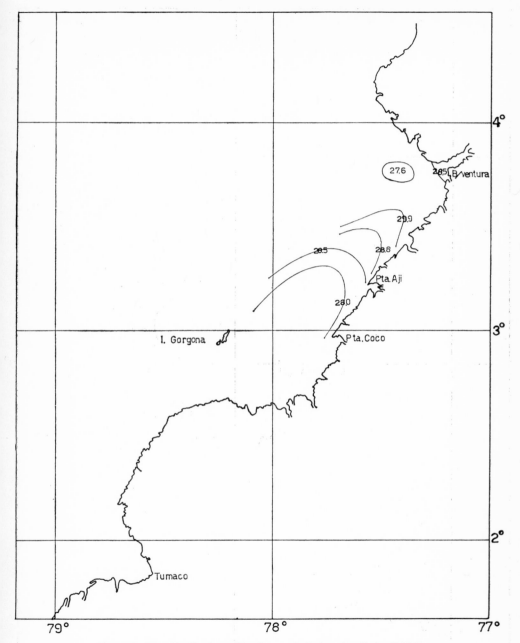

Figure 2 The distribution of isotherms during cruise C, April 1969

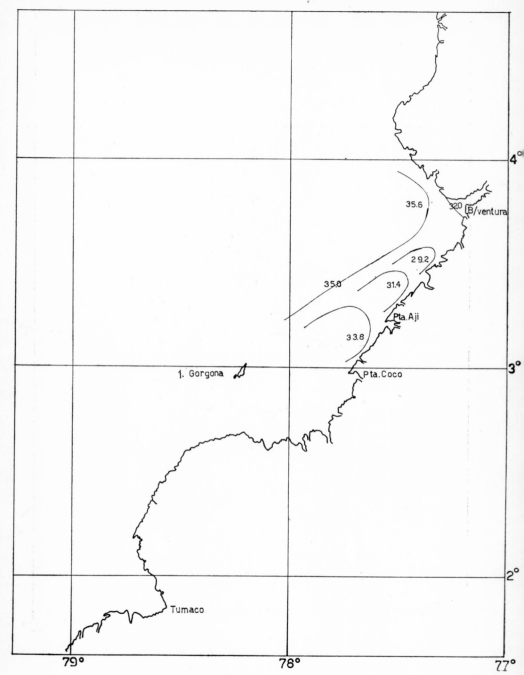

Figure 3 The distribution of isohalines during cruise C, April 1969

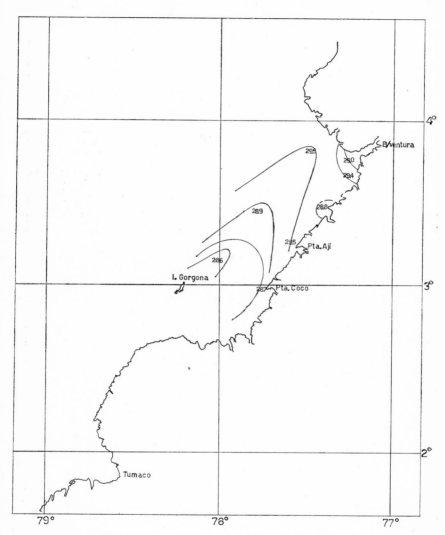

Figure 4 The distribution of temperature during cruise D, May 1969

1*

Fertility of the Sea

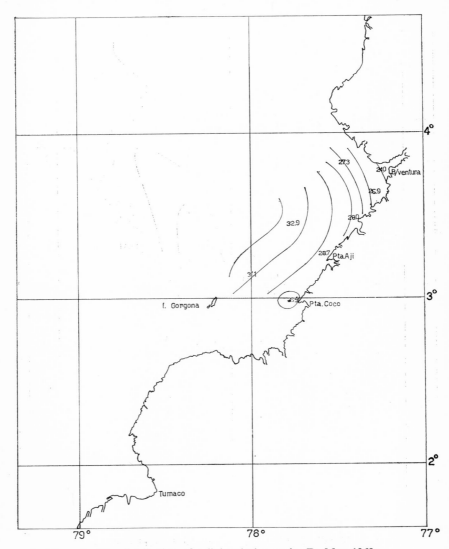

Figure 5 The distribution of salinity during cruise D, May 1969

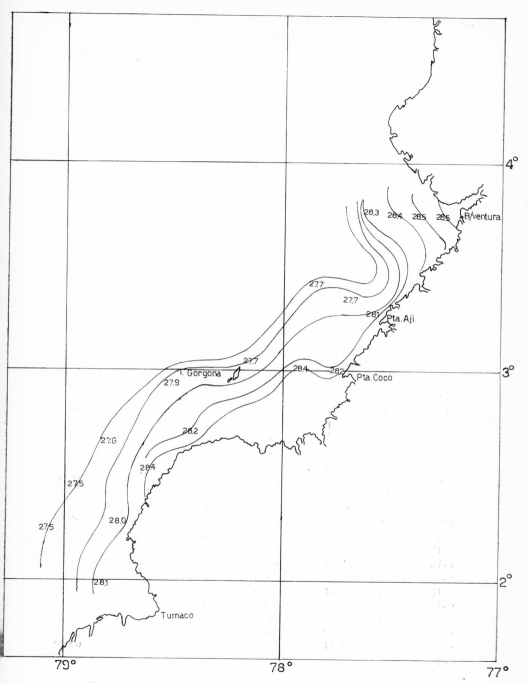

Figure 6 The distribution of isotherms during cruise F, July 1969

Fertility of the Sea

Table 1   Data of the cruises C, D, F, and G with results
indications about the presence of two groups

Station No.	Position Lat. N	Position Long. W	Date	Time	T°C sup.	Thermo Depth.
C-1	3° 46	77° 15	IV-4-69	11:00–11:15	28.54	
C-2	3 46	77 28	IV-4-69	14:15–14:30	27.63	
C-3	3 24	77 46	IV-7-69	09:15–09:30	28.50	
C-4	3 08	77 41	IV-7-69	11:45–12:00	28.07	
C-5	3 23	77 30	IV-7-69	13:20–13:35	28.89	
C-6	3 31	77 25	IV-7-69	15:00–15:15	29.99	
D-1	3 46	77 15	V-16-69	11:20–11:35	29.07	
D-2	3 32	77 18	V-16-69	13:00–13:15	29.42	
D-3	3 27	77 26	V-16-69	14:50–15:05	28.87	
D-4	3 15	77 37	V-16-69	16:25–16:40	29.56	
D-5	2 58	77 48	V-17-69	9:10–9:25	28.77	
D-6	3 08	78 02	V-17-69	10:10–10:25	28.67	
D-7	3 26	77 47	V-17-69	13:40–13:55	28.97	
D-8	3 48	77 29	V-17-69	15:45–16:00	29.52	
F-1	3 45	77 17	VII-1-69	5:30–5:45	28.55	30–40 m
F-2	3 45	77 24	VII-1-69	6:45–7:00	28.52	40–50
F-3	3 45	77 29	VII-1-69	8:45–9:00	28.47	40–50
F-4	3 46	77 36	VII-1-69	10:30–10:45	28.33	30–40
F-5	3 25	77 51	VII-1-69	14:25–14:40	27.71	20–30
F-6	3 21	77 41	VII-1-69	16:45–17:00	27.71	
F-7	3 16	77 37	VII-2-69	6:50–7:05	28.12	20–30
F-8	3 00	77 46	VII-2-69	8:50–9:05	28.28	
F-9	3 00	77 56	VII-2-69	11:00–11:15	28.47	
F-10	3 01	78 09	VII-2-69	12:50–13:05	27.77	40–50
F-11	2 43	78 26	VII-2-69	17:20–17:35	28.21	
F-12	2 29	78 37	VII-2-69	20:15–20:30	28.41	
F-13	2 14	78 45	VII-2-69	23:05–23:20	28.04	30–40
F-14	1 59	78 50	VII-3-69	1:45–2:00	28.10	
F-15	2 13	73 05	VII-3-69	5:30–5:45	27.51	
F-16	2 25	78 57	VII-3-69	7:50–8:05	27.50	
F-17	2 39	78 49	VII-3-69	9:45–10:00	27.68	
F-18	2 54	78 31	VII-3-69	13:20–13:35	27.93	
G-1	3 45	77 17	VII-11-69	4:30–4:45	27.50	
G-2	3 44	77 29	VII-11-69	6:15–6:30	27.50	
G-3	3 45	77 39	VII-11-69	8:20–8:35	27.47	
G-4	3 25	77 51	VIII-11-69	11:30–11:45	27.75	60–70
G-5	3 20	77 42	VIII-11-69	13:30–13:45	28.16	

of the countings on zooplankton aliquots, with
of Chaetognaths all over the four cruises

Thermo range	S^0/$_{00}$ surf.	O$_2$ surf.	No. org aliguot	No. org cc/15′	Vol.-plankton cc/15′	G. euneritica	G. serrato dentata
	32.016	6.55	12.128	86.63	45	×	×
	35.635	5.8	33.296	237.82	100	×	×
	35.027	6.65	4.584	32.74	50	×	
	33.808	6.30	11.948	85.34	50	×	×
	31.4053	5.00	10.976	78.40	60	×	
	29.2076	1.25	16.092	114.94	60	×	
	24.005	5.50	15.440	110.29	30	×	
	26.998	6.25	19.684	140.60	100	×	
	28.020	6.50	8.708	62.20	40	×	
	28.736	6.35	10.308	73.63	40	×	
	26.408	6.57	27.532	196.66	60	×	
	31.103	6.40	6.748	48.20	140	×	×
	32.947	6.10	6.808	48.63	30		×
	27.305	4.54	16.884	120.60	50	×	
8.50		6.50	13.000	92.86	100	×	
7.08	28.164	6.70	1.412	10.09	75	×	×
7.80	28.723	6.55	1.348	9.63	45	×	×
6.60	30.745	6.70	15.764	112.60	50	×	×
7.10	30.384	6.65	36.88	26.34	45	×	×
			35.252	251.80	50	×	×
8.30	29.120	6.60	34.552	246.80	40	×	×
	28.723	6.45	54.238	380.37	60	×	
	31.647	6.60	5.420	38.71	100	×	
8.40	30.709	6.55	32.332	230.94	40		×
	29.481	6.35	38.792	277.09	50		×
	30.510	6.40	49.680	354.86	50		×
6.10	29.752		62.936	449.54	50		×
	30.239		12.884	92.03	50		×
	35.889		15.176	108.40	70		×
	35.889	6.45	27.708	197.91	50		×
	32.496	5.80	8.408	60.06	10		×
	29.842	5.90	11.500	82.14	40		×
			27.064	193.31	80	×	
	27.278	6.20	5.980	42.71	40	×	
	31.830	7.05	4.484	32.03	25		×
6.60	31.299	2.40	5.700	40.71	30		×
	30.691	7.00	9.412	67.23	30	×	×

Fertility of the Sea

Table 1

Station No.	Position Lat. N	Long. W	Date	Time	T°C sup.	Thermo Depth.
G-6	3° 15	77° 36	VIII-11-69	15:10–15:25	28.05	
G-7	3 00	77 46	VIII-11-69	18:10–18:25	28.25	
G-8	3 00	77 55	VIII-11-69	20:05–20:20	28.22	
G-9	3 04	78 10	VIII-12-69	6:30–6:45	27.22	50–60
G-10	2 45	78 27	VIII-12-69	10:40–10:55	27.10	40–50
G-11	2 30	78 37	VIII-12-69	13:45–14:00	27.38	
G-12	2 14	78 45	VIII-12-69	16:25–16:40	27.40	50–60
G-13	1 59	78 49	VIII-13-69	7:10–7:25	26.93	50–60
G-14	2 13	79 05	VIII-13-69	8:45–9:00	27.20	
G-15	2 25	78 57	VIII-13-69	11:15–11:30	27.13	60–70
G-16	2 38	78 47	VIII-13-69	13:30–13:45	27.29	
G-17	2 54	78 30	VIII-13-69	16:35–16:55	27.43	40–50

be due to a decrease of the rivers run-off. In other words this month is marked by a progression to the coast of the equatorial pacific waters with a salinity of 35°/oo. The neritic water has a salinity of 28°/oo to 30°/oo. The transition water has a salinity range of 30°/oo to 35°/oo, with a plankton formed by the euryhaline organisms of both offshore and inshore planktonic communities.

August 11–13, 1969, Cruise G

The temperature distribution during this month shows the similar pheno-menon described above in cruise *F*: an intrusion of the equatorial-pacific water into the continental shelf, principally in the areas of the Golfo de Tortuga (Buenaventura) (Fig. 8), Golfo de Guapi and Tumaco Bay. This distribution of the isotherms has a remarkable influence on the primay production of the three sectors mentioned above.

The salinity during this month (Fig. 9) shows a noticeable decrease for the shore waters in Buenaventura and Tumaco sectors. In the Buenaventura area (Golfo Trotuga) the important run-off of the rivers San Juan and Raposo are the explanation of the surface salinity variations.

In the Gorgona Island and Guapi region we note an intrusion to the coast of the intermediate waters with salinity ranges of 30°/oo to 33°/oo. In the Tumaco area the isohalines of low concentration have been pushed out to

(*cont.*)

Thermo range	S⁰/₀₀ surf.	O₂ surf.	No. org aliquot	No. org cc/15′	Vol.- plankton cc/15′	G. euneritica	G. serrato dentata
	29.918	7.05	40.612	290.09	70	×	
	30.691	6.90	36.644	261.74	65	×	
	32.732	6.70	10.144	72.46	125	×	×
4.00	30.045	6.85	35.496	253.54	65		×
6.30	28.343	5.50	3.340	23.86	25	×	×
	28.307	5.25	20.408	145.77	60	×	×
6.00	27.841	5.50	27.984	199.89	75	×	×
6.80	28.198	6.67	12.488	89.20	50		×
	28.881	6.90	12.764	91.17	50	×	×
10.90	28.521	6.85	13.168	95.49	40	×	×
	31.783	6.05	63.084	450.70	50	×	×
7.40	32.140	8.02	15.564	111.17	35		×

the Banco de Tumaco (Fig. 1) at the station G-14, (cuadro 1). This difference between temperature and salinity distribution suggest that the isohalines are more useful to separate the water masses in the area. It seems too, that chaetognaths follow the distribution of isohalines better than temperature.

Oxygen, April 4–7, 1969, Cruise C

Oxygen seems to have a negligible role in zooplankton distribution off the Colombian Pacific Coast. Longhurst (1967) states that even the sub-tropical deficient-oxygen layer is not a limiting factor in the distribution of the zooplankton in Baja California. Longhurst (1967) suggests that the oxygen tension limit could be lower than 0.2 ml/l for the zooplankton off Baja California.

In the surface layers (where we have worked) it is generally accepted that an oxygen deficiency could be rarely found. In "red tides" however, the amount of oxygen available to the organism is generally lower.

In the region we have studied oxygen seems to have a small influence on the zooplankton distribution or relative zooplankton abundance. At least in this preliminary results there is no correlation between one and another.

In spite of this I will give a description of the fluctuations of this parameter for each of the four cruises we are analyzing.

28

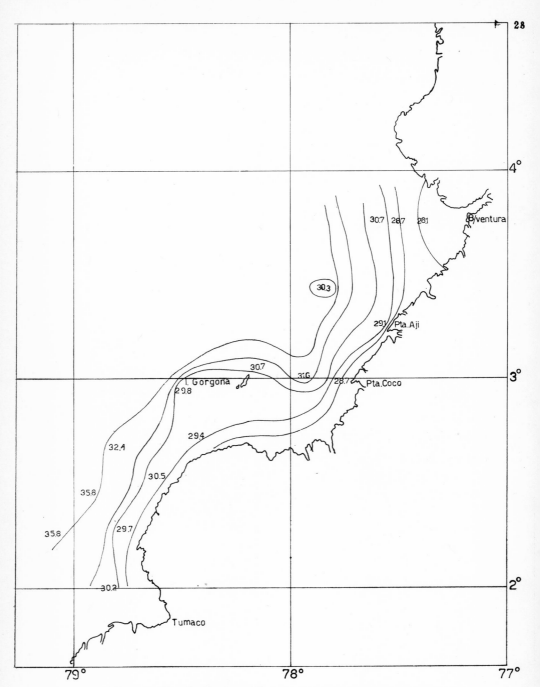

Figure 7 The distribution of isohalines during cruise F, July 1969

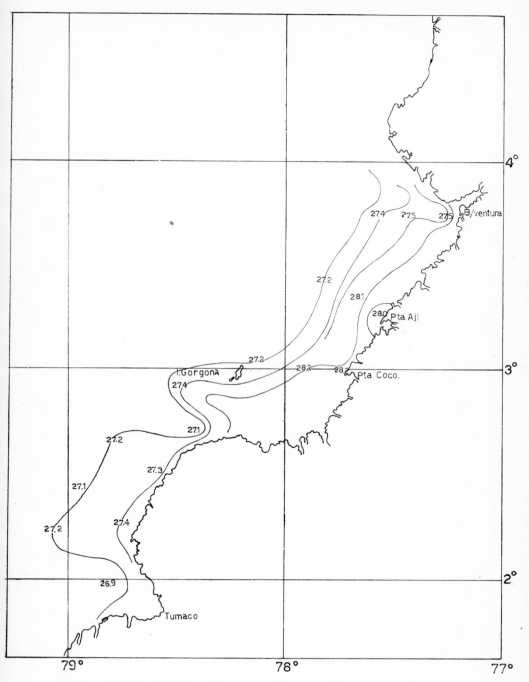

Figure 8 The distribution of temperature during cruise G, August 1969

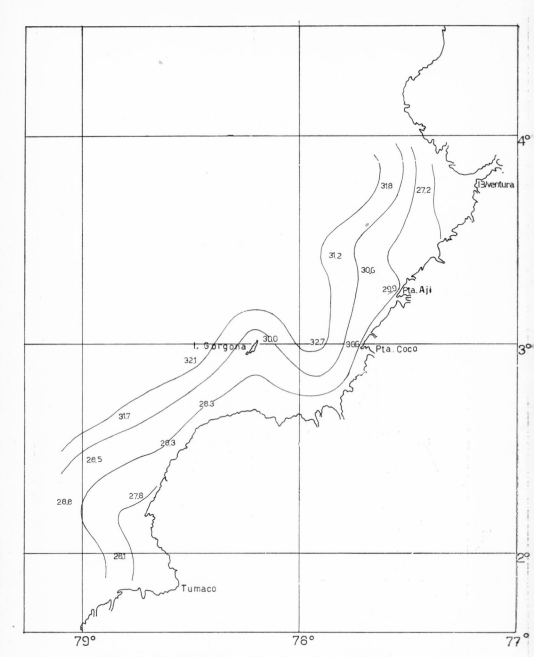

Figure 9 The distribution of isohalines during cruise G, August 1969

1. In April the oxygen shows a maximum at the station C-3 and a minimum at station C-6 (Fig. 1, Table 1). The station C-6 is situated on the mouth of the Cajambre River. This minimum oxygen concentration could be explained by the oxidation of the organic materials carried off by the river.

The value of the oxygen concentration for C-6 station is, however, 1.25 ml/l (Table 1), higher than the possible minimum value reported by Longhurst (1967) for Baja California.

2. The oxygen in May 16–17, 1969, shows a maximum at D-5 (6.57 ml/l) (Table 1) and a minimum at D-8 (4.54 ml/l). Any one of the values found on this cruise could not have a limiting role in zooplankton distribution. The water appears to have a high oxygen concentration at every station.

3. The oxygen during July 1969 has a range of 6.70 ml/l to 5.80 ml/l (cuadro 1). In other words, the waters have a high oxygen concentration over all the region. No correlation exists between zooplankton density and oxygen distribution. For example, the station F-1 (Fig. 11) with a high oxygen concentration (6.50 ml/l) has a population density per cubic centimeter equal to 92.86 organisms (Table 1).

Station F-10, with a similar oxygen concentration (6.55 ml/l) has a density of 230.94 organisms/cc./15′ that is to say, two-fold that of the F-1 station.

4. The oxygen in August 11–13, 1969 has a range of 7.05 ml/l at station G-3 (Fig. 1) and 2.40 ml/l at station G-4 (Table 1). There is no correlation between oxygen concentration and "standing crop". At the station G-6, we counted 290.09 organisms cc./15′, with an oxygen concentration of 7.05 ml/l, while at station G-3, which had the same concentration of oxygen (7.05 ml/l) we found the zooplankton relative abundance to be 32.03 organisms cc./15′.

Summarizing we will say that the oxygen concentration does not appear to be a limiting factor of the "standing stock" in the region surveyed. No correlation exists between zooplankton density and oxygen concentration. Finally, the amount of oxygen available to organisms in these surface waters is always high.

The chaetognaths and the water masses of the region

Many biologists in the last fifty years have shown the relationship between chaetognath distribution and water masses. Sund and Renner (1959) to cite only a recent paper, have studied the chaetognaths and their relationship to the equatorial and sub-tropical pacific waters. Bieri (1956) has shown the relation between chaetognaths and the incursion to the south of the tropical waters off Peru (the Corriente del Niño). Thereafter the same author (Bieri, 1959) studied the relationship between chaetognaths and water masses in the Pacific. Sund (1961) studying the chaetognaths collected between California and the Costa Rica Domo, grouped them in three categories according to their possible utilization as water mass indicators.

In the present work we considered the water masses present in the sector between Buenaventura, Gorgona Island and Tumaco Bay (Fig. 1). From the description I have given in the first part of this paper of the temperature and salinity conditions, we have concluded that there are in the region three water masses:

a) "*diluted water of the continental shelf*": marked by a salinity range variation of 28 to 30⁰/₀₀, and a temperature range variation of 28 to 29°C. The salinity variations are induced by river run-off. These rivers contribute to the near sea an important amount of nutrients, which is used in development by the zooplankton.

b) "*transition water type*": is an intermediate kind of water between continental shelf water and oceanic-pacific water. The salinity of this water mass fluctuates between 30 to 35⁰/₀₀ and the temperature varies from 28 to 27°C.

The plankton of this water primarily includes species with a wide tolerance to the salinity fluctuations.

c) *equatorial-pacific water*: the origin of this water is the Equatorial Counter-Current which flows West-to-East furnishing the off-shore water off Colombia, Panama and part of Central America. Near the Colombian coast the equatorial-pacific water has a salinity of 35⁰/₀₀ and a temperature of 27°C. The plankton of this water is quite different from the plankton of the continental shelf water.

A taxonomic and statistical analysis of plankton communities is a very difficult task and with special knowledge requirements. We do not intend to undertake such a study at the present time. We are attempting, however, a morphological study of the chaetognaths in the coastal waters off Colombia. Initially we have defined two groups followed in our samples. These are:

1. *"Euneritica Group"*

Characteristics of this group are the species similar to *Sagitta setosa*, which is typically a neritic chaetognath. This group includes *S. euneritica*, *S. friderici*, *S. tenuis* and *S. peruviana* together with the species which have a wide distribution because they can tolerate large variations in salinity. Such species are *S. neglecta* and *S. enflata*. These species are found in neritic waters *but in small numbers*.

2. *"Serratodentata Group"*

These species are distributed in equatorial-pacific waters. Following the distribution of these species I hope to know how this water mass fluctuated in this region. Such species are *S. pacifica*, *S. serratodentata*, *S. pseudo-serratodenta*, *S. pulchra* and *Pterosagitta draco*.

The species which have a wide spread distribution include *S. enflata*, *S. neglecta*, *S. nagae* and *S. robusta*.

The presence of this group is indicated in Table 1.

Chaetognaths distribution during C. Cruise

The station C-1, is marked by the presence of "eunezitic" chaetognaths (Fig. 1) and few individuals of the *"serrato-dentata"* group.

The station C-2 is characterized by a dominance of individuals of "serrato-dentata" group and a few individuals of "eunezitic" group. In both stations the wide spread species are well represented.

If we look at the isohaline distribution we can see that at the station C-2 and C-3 there is an intrusion of the $35^o/_{oo}$ isohaline (equatorial-pacific water) toward the coast. So, the station C-2 is located at the place where the two water masses mix. Station C-4 presents the same characteristic as C-2 and the presence of the $33^o/_{oo}$ isohaline is the explanation for this phenomenon. Stations C-5 and C-6, located at the rio Cajambre and rio Jurumanguí mouths, are typically nezitic, and are characterized by the presence of "eunezitic" species.

Chaetognaths distribution in D

Stations D-1, 2, 3, 4, 5, and 8 are characterized by the presence of "eunezitic" species. All these stations are located in the coastal area at the rivers mouths. Stations D-6 and D-7 are characterized by the presence of wide spread or euryhaline species. Looking at the isohaline distribution (Fig. 5) we note that stations D-1, 2, 3, 4, 5 and 8 are located in the diluted water while stations D-6 and D-7 are located in the transition water.

Chaetognaths distribution in F

The stations F-1, 8 and 9, located in diluted waters, are characterized by the presence of the "eunezitic" group. Stations F-2, 3, 4, 5, 6 and 7 are characterized by the presence of wide spread species, which follows the isohaline distribution (Fig. 7). At stations F-10, 11, 12, 13, 14, 15, 16, 17 and 18 we observe the presence of the "serratodentata" group represented by a small number of individuals mixed with the wide spread species. Looking at the isohalines (Fig. 7) we observe that values corresponding to these stations could be classified in the transition water, except for stations F-15, 16 and 17 located in oceanic waters.

Chaetognaths distribution in G

Stations G-1, 6 and 7 are marked by the present of "eunezitic" chaetognaths.

Stations G-2, 5, 8, 10, 11, 12, 14, 15 and 16, located in the transition waters, are characterized by the wide spread species. The salinity values for these stations (Fig. 9) shows that we have in this sector transition water. Stations marked by the "serratodentata" group (G-3, 4, 9 and 17) are located in the areas where there is an intrusion toward the continental shelf of the oceanic water (Fig. 9). It shows that the chaetognath distribution in this area follows the position of the isohalines.

Relative abundance of zooplankton

The total volumes obtained for each of the stations examined are included in Table 1, including the organism number per aliquot (140 c.c. of plankton sample diluted to 500 c.c.) and the index abundance per cubic centimeter 15° tow.

Analyzing the total plankton volume collected and the index of density per cubic centimeter we observe that there is not always a positive correlation. That is to say that a high plankton volume collected does not always correspond to a high organism density. The explanation of this phenomenon is that in the plankton samples taken in the "oceanic" region there is a high volume of organisms like Medusas which could increase abnormally the total plankton volume. In the samples from coastal waters, where there is a dominance of small specimens like chaetognaths, Copepodes and Ostracodes, there is a high plankton density per cubic centimeter.

In other words, there is a higher density of organism per cubic centimeter at coastal stations than in oceanic stations.

The zooplankton in April 4–7, 1969

During the cruise C in April 1969 (Fig. 10) the stations C-1, 5, 4 and 2 show a low density of organisms per cubic centimeter. We found a high density at station C-2 (237.82 organisms/cc./15′) which is explained by the contact of the two water masses (oceanic and neritic) at this point (Fig. 3).

The zooplankton in May 16–17, 1969

During this month we observed that the stations located near the coast (Fig. 11) present a high density of organism per cubic centimeter. Such stations are D-1, 2 and 5. These high concentrations of zooplankton in coastal waters seems to be in relation to the position of oceanic waters off the coast line and the increasing of river run which is important during this month. This run-off possibly produces a nutrient enrichment of the coastal area.

The zooplankton in July 1–3, 1969

During this month we observed that the coastal stations (Fig. 12), (F-6, 7, 8, 11, 12 and 13), have a high zooplankton density. The areas near Punta Aji and Punta Coco present a standing stock with a range density of 246.80 organisms/cc./15′; (F-7) and 380.37 organisms/cc./15′ at F-8 (Table 1). In the Gorgona Island we found 230.94 organisms/cc./15′; (F-10). At station F-11, (Punta Guascama) we observed 277.09 organisms/cc./15′ and 449.54 organisms/cc./15′ at the Patia River mouth. In other words, stations located in the neritic area present the highest population density during this cruise.

The zooplankton in August 11–13, 1969

The zooplankton density distribution is almost the same as during July. The most productive sector seems to be Punta Aji and Punta Coco, which present values of 290.09 organisms/cc./15′ (Table 1, Fig. 13) for station G-6 and 261.74 organisms/cc./15′ at G-7. The Gorgona Island has a value of 253.54 organisms/cc./15′ (G-9).

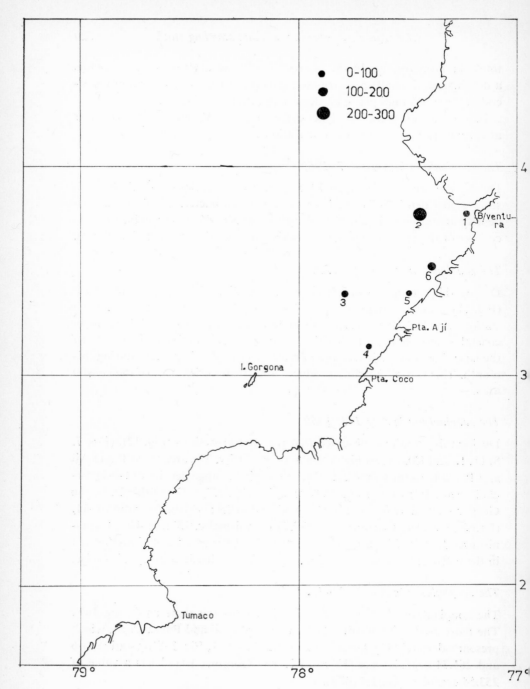

Figure 10 Zooplankton density from cruise C, April 1969

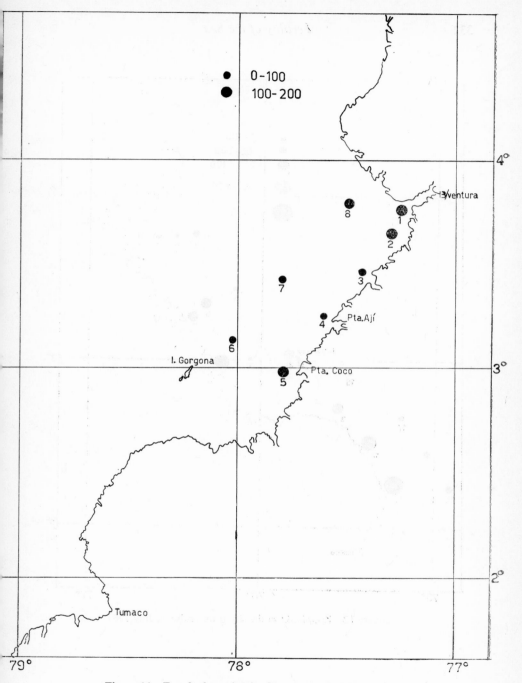

Figure 11 Zooplankton density from cruise D, May 1969

2*

Fertility of the Sea

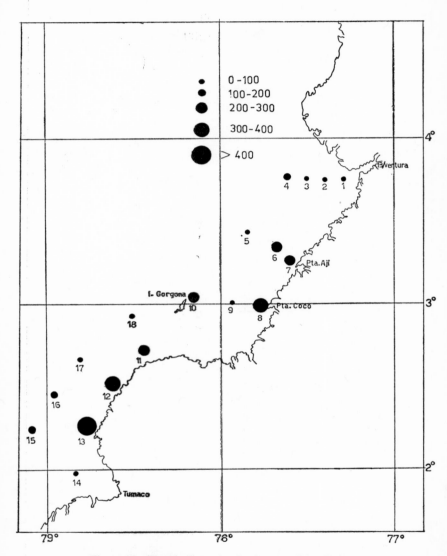

Figure 12 Zooplankton density from cruise F, July 1969

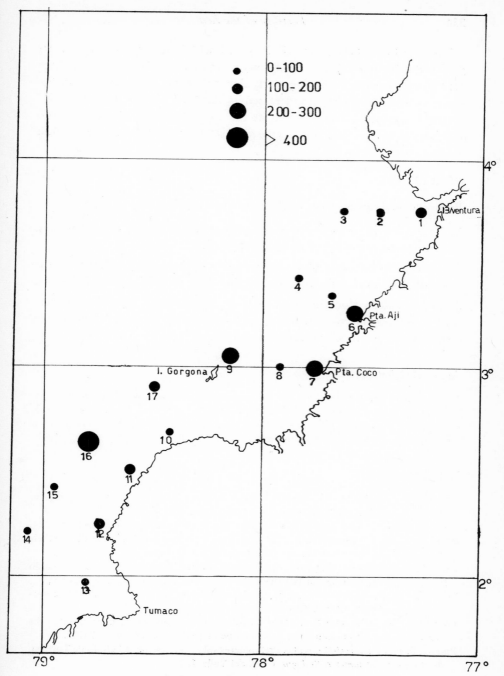

Figure 13 Zooplankton density from cruise G, August 1969

334 *Fertility of the Sea*

The area Punta Guascama-Rio Patia still with a high density (145.77 organisms/cc./15′ (G-11) and 199.89 organisms/cc./15′ (G-12)). These three sectors appear with a standing stock higher than the standing stock of the Golfo de Tortuga, which has increased its value to 193.31 organisms/cc./15′ (G-1). The station G-16, located at the contact surface between coastal water and oceanic water, presents the highest value, 450.70 organisms/cc./15′.

CONCLUSION

From the analyses we have made it could be possible to conclude:

1.   There are three different water masses in the sector: coastal waters, transition waters and equatorial-pacific waters.

2.   It seems possible to follow the evolution of these three waters from the study of chaetognath distribution by using them as hydrological indicators.

3.   The standing stock seems to be more important in the coastal waters than in the oceanic waters. It may be explained by the enrichment of the coastal waters by a permanent river run-off.

Acknowledgements

I would like to thank the directors of the Universidad del Valle, especially Drs. Enrique Tomo, vice-rector, Ramiro Tobon, dean of the Division of Sciences and Jose I. Borrero, chairman of the Biology department. All of them have made possible this investigation and helped the author with invaluable advice.

References

BENNETT, EDWARD B. (1966). Monthly Charts of Surface Salinity in the eastern tropical Pacific Ocean. *Int. Am. Trop. Tuna Comm.* **11**, No. 1.
BIERI, R. (1956). The Chaetognatha Fauna off Peru in 1941. *Pacific Science.* XI, 256–264.
BIERI, R. (1959). The Distribution of the Planktonic Chaetognatha in the Pacific and their Relationship to the Water Masses. *Limnology & Oceanography.* Vol. **IV**, No. 1.
FAGETTI, E. and FISCHER, W. (1964). Resultados Cuantitativos del Zooplankton Colectado frente a la Costa Chilena por la Expedición "Marchile I". *Montemar.* **4**, 137–200.
FURNESTIN, M. L., MAURIN, C., LEE, J. Y. et RAIMBAULT, R. (1966). *Elements de Planctonologie Appliquée*, I.S.T.P.M., Paris.
LONGHURST, ALAN R. (1967). Vertical Distribution of Zooplankton in relation to the eastern Pacific oxygen minimum. *Contributions of Scripps Ins. Oceang.* **37**, 267.
PINEDA, P. FRANCISCO (1958). Notas Preliminares sobre el plancton del Golfo de Tortuga. *Boletin del Departamento Biologia*, Univ. del Valle. **I**, 2.

PINEDA, P. FRANCISCO (1969). Los Chaetognatos de la Expedition Oceanografica No. XX de la Universidad de Stanford. *Boletin del Departamento Biologia*, Univ. del Valle. **2**, 1.

SUND, PAUL N. (1961). Some features of the Autoecology and Distributions of Chaetognatha in the Eastern Tropical Pacific. *Int. Am. Trop. Tuna Comm. Bull.* **V**, 4.

SUND, P. N. and RENNER, J. (1959). The Chaetognatha of the EASTROPIC EXPEDITION, with notes as to their possible value as indicators of hydrographic conditions. *Bull. Int. Am. Trop. Tuna Comm.* **III**, 9.

Distribution of nutrients in the sea and the oceanic nutrient cycle

H. POSTMA

Netherlands Institute for Sea Research, Texel
Norntje, Postbus 59, Holanda

Abstract

The general distribution of nutrients, specifically phosphate, in the oceans is described and a simple quantitative model is developed to indicate the main pathways of transport. A general estimate is made of the rate of nutrient supply to the upper water layers and of the ratio between supply by vertical mixing and by upwelling. The relationship between the distribution of nutrients and of primary production is discussed; recycling of phosphate in the warm surface layer seems to be of the order of 50%.

Resumen

E descrita a distribuição geral de nutrientes nos oceanos especìficamente fosfato, e um modêlo quantitativo simples foi desenvolvido para indicar os principais caminhos de transporte. É feita uma estimativa geral da razão de suprimento de nutrientes para as camadas superiores e da razão entre suprimento por mistura vertical e ressurgência. As interrelações entre a distribuição de nutrientes e produção primária são discutidas, a reciclagem de fosfato para a camada quente superior parece ser da ordem de 50%.

GENERAL DESCRIPTION

The distribution of nutrients in the oceans is determinded by the oceanic circulation, biological processes of uptake and mineralization, by the settling of organic debris through the water column and subsequent regeneration of nutrients, by migration of animals and by supply from the land. The last factor is only of minor importance in the open sea.

The general characteristics have been well established now for many years. A large amount of information was already collected in the years between 1920 and 1940 by a number of ocean expeditions. This material has been greatly extended in recent years.

One of the major problems of nutrient research is that of the supply of nutrients to surface waters. Most of the nutrients, such as phosphates, nitrates, silicates and other minor elements of biological importance are concentrated in deeper waters outside the productive zone. This is essentially due to the fact that plankton organisms and their remains, although very small, sink slowly downward. In addition, nutrients are carried down by vertical migration of animals. They return to the surface by vertical mixing and upwelling, processes which are much more active in one part of the ocean than in another, causing one region to be much richer in nutrients than another.

The rate of supply of nutrients, together with light, determines to a large degree the biological productivity of the oceans. Generally speaking, these two factors rarely support each other. Where light is abundant, as in tropical and subtropical waters, nutrient supply is low as a result of the high temperature and the resulting great stability of the surface layer. In temperate and polar regions, on the contrary, vertical mixing is very intensive during part of the year, but illumination is poor. The most fertile parts of the oceans are thus upwelling areas in tropical waters. These areas are chiefly restricted to the equatorial zone and the eastern borders of the oceans (Wooster and Reid, 1963).

In the deep sea, the concentration of a nutrient substance at a specific point is determined by the content of organic remains at that place, their rate of mineralisation and by the rate of water renewal. There are indications that the rate of mineralisation of phosphates and nitrates at first decreases exponentially with depth from the surface downward (Riley, 1951). In deeper waters, below about 1000 metres, the distribution of suspended and dissolved organic matter seems very uniform (Duursma, 1961; Menzel, 1967), so that these vertical differences in the rate of mineralisation are probably very small.

Water renewal is relatively rapid in bottom waters, because of the supply of cold surface water, chiefly from Antarctic regions. In combination, these processes lead to the formation of phosphate and nitrate maxima, and oxygen minima, in intermediate waters. This does not hold, for example, for silicate and calcium carbonate, which mostly show a gradual increase in concentration down to the ocean floor, due to the fact that these substances go into solution in abyssal waters or partly even enter the water from the ocean bottom itself.

The oxygen minimum is in most areas located in a higher level than the phosphate and nitrate maximum, especially in tropical and subtropical regions. Here oxygen is not only used by the oxidation of organic matter, but part of it escapes to the atmosphere (Eriksson, 1959). The latter process reduces the depth of the oxygen minimum layer (Postma, 1964).

The nutrient contents in deep water of the Pacific and Indian Oceans north of the Equator are a factor three higher than in the Atlantic Ocean. On the northern hemisphere, deep water is only formed in the Atlantic. The deep and bottom water of this part of the ocean is, therefore, renewed relatively fast. In addition, the over-all water circulation in the North Atlantic Ocean is anti-estuarine (Refield, Ketchum and Richards, 1963). Surface and sub-surface water is carried north over the Equator, whereas deep water returns to the south. This type of water circulation depresses nutrient concentrations. The North Pacific, on the contrary, has an estuarine type of circulation, which forms a trap for nutrients.

This paper discusses in a semi-quantitative way some of the processes which lead to the described nutrient distribution. The discussion will be based on the distribution of phosphate about which data are most abundant. Besides phosphate, data on oxygen consumption and primary production will be used. It is, therefore, necessary to discuss first the ratio of change of concentration of phosphorus, oxygen and carbon in biological processes in the sea.

Atomic ratios

The proportions in which oxygen, carbon dioxide and nutrients change in a certain water mass is determined by the chemical composition of organic matter. The composition of oceanic plankton seems to be surprisingly uniform, so that it has been possible to determine atomic ratios of change which hold for most oceanic areas. The most accepted formula is $\Delta P : \Delta N : \Delta C : \Delta O = 1 : 16 : 106 : -276$ by atoms. The $\Delta P : \Delta N$ ratio has been determined both from the composition of plankton and of sea water; the other ratios from the composition of plankton only (Redfield, Ketchum and Richards, 1963).

A main difficulty in determining these ratios from sea water is that oxygen and carbon dioxide, in addition to being used in the biological cycle, are also exchanged with the atmosphere. As a result, oxygen utilisation by decaying organic matter can easily be overestimated in surface and subsurface water of tropical and subtropical areas. The influence of escape to the atmosphere is important down to the level of the phosphate (or nitrate) maximum, as will be shown later.

Hence, in order to exclude the immediate influence of air-sea interchange, only data from below this level should be considered. These data give straight line relationships. Figure 1 shows these relations for the Pacific and Atlantic Oceans, based on information from a large number of expeditions. The two lines are parallel; the Pacific line originates from Antarctic surface water with a phosphate content of 1.6 μg-at/l and an oxygen content of 7.2 ml/l; the Atlantic line from North Atlantic surface water with values

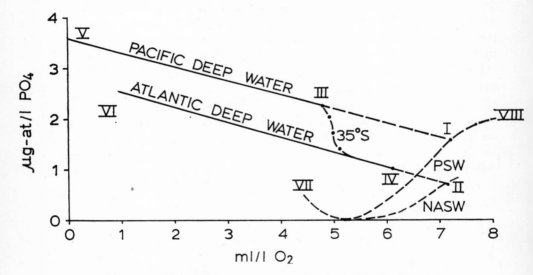

Figure 1 Relation between oxygen and phosphate of Pacific Deep Water (III–V) and north and central Atlantic Deep Water (IV–VI) in depths below the phosphate maximum. Simplified relation for Pacific surface water (PSW VII–VIII) and North Atlantic surface water (NASW VII–II) along north-south sections.

I: source of Antarctic Bottom Water; II: source of North Atlantic Deep Water; III–IV: transition zone between the two main lines in the southern Atlantic around 35°S; the connecting line refers to values from the deep water salinity maximum; V: phosphate maximum and oxygen minimum in the eastern Pacific; VI: phosphate maximum and oxygen minimum in the eastern Atlantic; VII: equatorial phosphate maximum; VIII: phosphate maximum of ascending deep water in the Antarctic of 0.8 μg-at/l and 7.2 ml/l. The lower position of the Atlantic line is caused by factors discussed in the first chapter. The connection between the two main lines is found in the South Atlantic Ocean, where the two water types are mixing. In the figure, this connection is represented by the P—O_2 relation of the core of Atlantic Deep Water.

The slope of the lines corresponds with a phosphate regeneration of 0.27 μg-at/l against an oxygen consumption of 1 ml/l, or an atomic ratio $\Delta P : \Delta O_2 = 1 : -336$. Comparable data on carbon dioxide (Postma, 1964) indicate that the ΔO_2—ΔCO_2 relation does not differ from that used in the above formula, so that the corrected formula, on the basis of deep water observations, is $\Delta P : \Delta C : \Delta O = 1 : 138 : -336$ by atoms. These ratios will be used in the calculations of this paper, keeping in mind that they may not correspond with the average chemical composition of plankton. It seems possible that phosphorus is liberated from organic compounds more rapidly than the rate of oxidation of the organic matter itself.

Examples of P—O_2-diagrams from the surface into deep water at stations in the North Atlantic and Pacific Ocean are shown in Figure 2. The relationship of values from above the phosphate maximum deviates from the deep

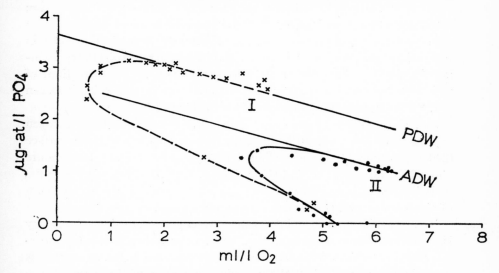

Figure 2 Relation between oxygen and phosphate along vertical.

I: Station Downwind 7; 7° 08′ N; 130° 00′ W (Scripps Inst. Oceanogr. exp. 1957–1958); II: Station Casco (Equalant II); 35° 32′ N; 53° 10′ W. PDW is relation for Pacific Deep Water; ADW for Atlantic Deep Water

water line. The horizontal distance to the main line indicates the amount of oxygen lost to the atmosphere. Diagrams of this type show that losses mainly occur below the warm part of the surface layer. This indicates that, notwithstanding the great stability of this warm layer, vertical mixing through this layer is of great importance. This conclusion agrees with the fact that the salinity in the warm water layer is determined by net evaporation and the average value of the salinity at a depth of 400–600 metres (Sverdrup *et al.*, 1946).

Oceanic transport of nutrients

The relationship between the oceanic circulation and biological productivity, summarily described in the preceding paragraph, can be evaluated more precisely if transport rates of nutrients from one part of the ocean to the other are better known. A complete description of nutrient circulation obviously presupposes a complete quantitative knowledge of ocean current and mixing systems, which is not available. However, a rough quantitative picture can be obtained by the use of a simplified model.

In this model the ocean is divided into a few large reservoirs and the amounts of water and nutrient moving from one reservoir to the other are determined. Phosphate is again selected as the nutrient about which the most abundant information is available. The elements of the model are shown in Figure 3 A. Essentially, the model consists of a central reservoir (I), representing the circumpolar water of the Southern Ocean, the northern limit of which has been placed at 40°S; reservoirs representing the Atlantic Ocean (VI, VII, VIII and IX) and the combined Pacific and Indian Oceans.

Nrs. V and VI represent warm surface water, which is found between approximately 40°S and 40°N and covers about 75% of the ocean surface (240 × 10^{12} m^2). The average salinity is 35.3°/oo and the average temperature about 25°C (Sverdrup *et al.*, 1946; Eriksson, 1959). The boundary with the underlying water mass is formed by a sharp thermocline at an average depth of about 100 m. Nr. II represents cold surface water.

Nrs. IV and VII represent intermediate water masses; the lower boundary is placed in the phosphate maximum. Thus, water exchange with the underlying deep water does not cause a net phosphate transport. Nr. IX represents Atlantic water north of 40°N. The cold water mass of the North Pacific Ocean, which plays only a minor role in the general ocean circulation, is not included in the model. Nrs. III and VI represent the deep water below the phosphate maximum.

Transport of phosphate from one reservoir to the other takes place in three ways: (1) by advention, (2) by sinking of organic matter and (3) by vertical mixing. The main trajectories of water movement are shown in Figure 3B. Sinking of organic matter carries phosphorus from the surface waters into deeper layers. Vertical mixing moves phosphorus upward.

A steady state will be assumed, so that the amounts of phosphate entering and leaving each reservoir will be the same. As a result, relatively few numerical data are needed to calculate these amounts. Transport values of vertical mixing will be derived from differences in the balance.

a Bottom water is formed in the Antarctic (transport from II to I) at a rate of 0.8×10^{15} m³/year (Munk, 1966). The phosphate content of this water mass is represented by the value at the point of intersection between the deep water line and the surface line in the P—O_2 diagram of Figure 2; it amounts to 1.6 µg-at/l. Since the volume of the Atlantic Ocean is about one third of that of the other oceans combined, it is assumed that the Bottom Water is also divided over the oceans in this proportion.

b Deep water is formed in the North Atlantic Ocean at a rate of 0.2×10^{15} m³/year (Sverdrup *et al.*, 1946). This water has an average phosphate content of 0.8 µg-at/l (Fig. 2). After sinking, it is carried north through reservoir VI as North Atlantic Deep Water.

c Net transport of phosphate into the warm surface layer (IV to V and VII to VIII) can be calculated by means of equations developed by Wüst (1936), Eriksson (1959) and Montgomery (1959). These equations are based on the rate of water displacement in this layer as derived from net evaporation (40 cm/year) and the salinity difference between the water in the layer and the underlying water (35.3⁰/₀₀ against 34.6⁰/₀₀). The difference in phosphate concentrations is estimated at 1 µg-at/l. The calculation gives a net upward transport of 20 µg-at/m²/year. Over the whole warm water layer this amounts to 5×10^{12} g-at/year.

d The liberation of phosphorus from organic remains in the deep water (reservoirs III and VI; total volume 10^{18} m³) can be derived from the oxygen consumption in abyssal water, which has been determined by a number of authors (Riley, 1951; Postma, 1958; Munk, 1966). An average value of 0.002 ml/l/year will be used here. The biochemical ratio of phosphorus and oxygen has been discussed on p. 6. Total phosphorus supply to abyssal water thus comes at 0.5×10^{12} g-at P/year.

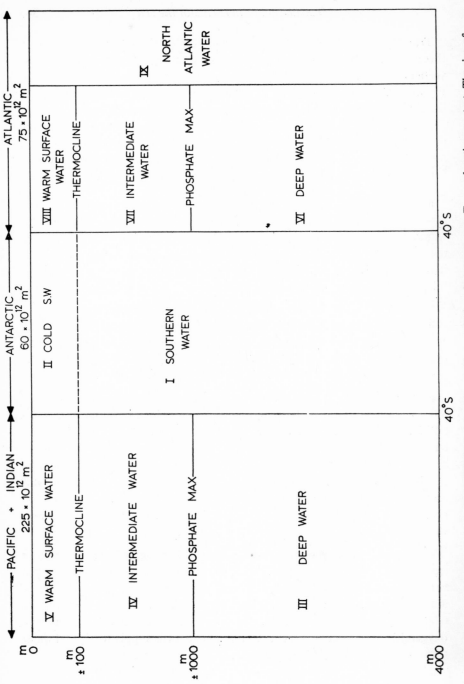

Figure 3A Simplified ocean model for calculations of net water and phosphate transport. For explanation see text. The size of

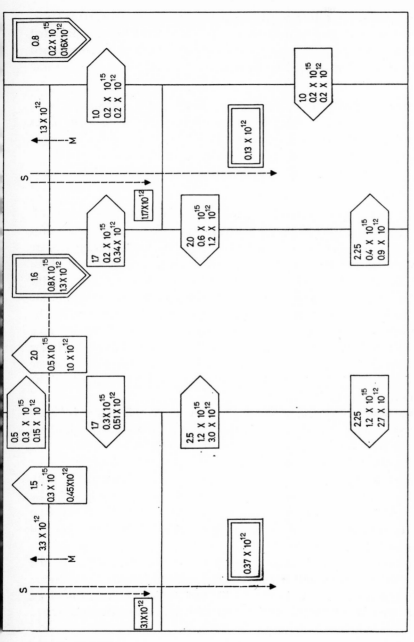

Figure 3B Net transport of water and phosphate between reservoirs calculated on the basis of data discussed in the text. Every arrow gives, from top to bottom, the phosphate concentration, in μg-at/l; the net amount of water, in m³/year, and the net amount of phosphate transferred, in g-at/year. S indicates transport by sinking of organic matter; M transport by mixing. A steady state is assumed, in which the amounts of water and phosphate entering and leaving each reservoir are equal. Small differences in the balance are caused by rounding off of numbers

Obviously, since 5×10^{12} g-at P/year returns to the warm surface layer, it must be assumed that the difference, 4.5×10^{12} g-at P/year, is liberated from sinking organic matter in reservoirs IV and VII. The amounts mentioned under c and d will be divided in the ratio 1 : 3 over the two main oceanic systems.

e Besides the phosphate concentrations already mentioned under a and c the following values are used: for water leaving the warm water layer 0.5 µg-at/m³; for intermediate water at 40°S 1.7 µg-at/m³; for bottom water at the same latitude 2.25 µg-at/m³; for Pacific Deep Water returning to the Antarctic 2.50 µg-at/m³ and the corresponding Atlantic Deep Water 2.0 µg-at/m³.

The above data, combined, with the assumption of steady state conditions, are sufficient for a calculation of the transport rates of water and phosphorus and leave no choice of a realistic alternative. An interesting feature is that the Antarctic Bottom Water, in flowing north, takes with it about the same amount of deep water. The mixture returns in higher layers. In the Pacific, this is enriched with the phosphate sinking into deep water; in the Atlantic this enrichment also occurs, but is more than compensated for by the addition of relatively poor North Atlantic Water. Part of the returning water, which is slightly warmed up (Deacon, 1937), rises to the surface in the Antarctic region. Essentially, the phosphate sinking into deep water from the warm surface layer is carried back into this layer via the Antarctic and the Antarctic Intermediate Water.

The total flow of Intermediate Water must be considerable (0.5×10^{15} m³/year) to accomplish this transport; in the Atlantic, moreover, Intermediate water is the main feeding source of the North Atlantic Water. In the Pacific, part of the Intermediate Water returns south in the surface layer (the same may happen in the Atlantic, but this cannot be derived from the model). This component may be considered part of the upwelling system in the warm surface layer, the more so since the Intermediate water tends to follow the path of boundary currents of this layer (Wyrtki, 1967). This point will be discussed below.

The model further illustrates the fact that the circulation in the Atlantic Ocean is of an anti-estuarine character, which leads to a decrease in nutrient content, whereas in the Pacific and Indian Oceans it is of an estuarine type which causes enrichment with nutrients (Redfield et al., 1963).

The surface layers

According to the above calculations the warm layer receives 5×10^{12} g-at of phosphate per year, or 20 mg-at/m^2/year. This amount may be compared with the quantity utilized in photo-synthetic assimilation.

Primary production in tropical and subtropical areas varies from an average of 150 g/m^2/year of carbon in upwelling areas (Cushing, 1969) to 30 g/m^2/year in the large subtropical gyres (Ryther and Menzel, 1961). The average is 60 g/m^2/year. This corresponds with 36 mg-at/m^2/year of phosphorus or about twice the amount supplied from deeper layers. Thus, every phosphate molecule is on an average used twice for photosynthesis before it is lost by sinking; in other words 55% of the phosphorus used annually is supplied by vertical transport. This value may be compared with that of about 75% found in temperate waters (Ketchum, 1947 and Steele, 1957) which are, however, stable only part of the year. Recycling of nutrients may be more important in permanently stratified regions (Anderson and Banse, 1961).

The supply of phosphate from deeper water layers takes place by vertical mixing and upwelling. The relative importance of these two processes determines the pattern of phosphate distribution over a large part of the ocean surface. Wooster and Reid (1963) discussed the magnitude of upwelling in eastern boundary currents. This type of upwelling is estimated to occur over a zone of 12,000 miles with an average ascending velocity of 50 m per month over a width of 50 km. The volume of ascending water thus amounts to 0.6×10^{15} m^3/year. Assuming that the upwelling water contains 1 μg-at of phosphorus per litre, total supply equals 0.6×10^{12} g-at/year. From Cushing's review (lgbg) one might guess that twice this quantity may be contributed by upwelling in equatorial regions, to that upwelling in all would contribute about 1.8×10^{12} g-at/year or 30% of the overall supply of 5×10^{12} g-at/year.

The enriching influence of upwelling must be felt over very large distances away from the source of ascending water. A column of 1 m^3 in the warm surface layer filled with new deeper water contains about 100 mg-at of phosphate. With a recycling factor of 50%, this is sufficient for the primary production of 330 grams of carbon. At a production rate of the order of 100 grams of carbon per m^2 per year, depletion takes more than 3 years. In that period the water will be carried away from its source over large distances. Maps of surface phosphate distribution (for example Wattenberg, 1934 and

3*

Reid, 1962) show that values above 0.25 µg-at/l are found over about 30%
of the warm surface layer.

It follows from the above estimates that phosphate supply by vertical
mixing alone is 3.2×10^{12} g-at/year or 14 mg-at/m²/year. With a recycling
factor of 50% this amount is able to sustain a carbon production of
38 g/m²/year, which indeed corresponds with the production in areas beyond
the reach of upwelling water.

According to Figure 3B, 0.3×10^{15} m³/year or if the Atlantic Ocean is
included perhaps 0.4×10^{15} m³/year, of ascending water is carried away
into cold surface water. Net evaporation could account for another 0.1
$\times 10^{15}$ m³/year. In all, this is only 25–30% of all upwelling water. The greater
part of it or about 0.1×10^{15} m³/year is returned by vertical mixing to the
reservoir below the warm surface layer where it came from. Therefore,
upwelling and vertical mixing must in some way or other be closely connected.
It seems possible, for example, that the intensity of upwelling is one of the
factors regulating the extensiveness of the subtropical convergence regions
between 30° and 40° north and south. In this manner, besides increasing the
fertility in certain parts of tropical and subtropical ocean areas, upwelling
would indirectly cause a decrease in other parts. The over-all result would be
an augmentation of productivity differences between various areas of the
warm water layer.

North and south of the warm water region, where surface water masses
are generally unstable at least part of the year, nutrient supply is mostly
plentiful, but other factors, such as light and vertical mixing, restrict
productivity. Recent measurements show that over large parts of the
Southern Ocean primary production is of the same order of magnitude as in
subtropical regions ($\pm$ 30 g C/m²/year; Burkholder *et al.*, 1967). Even lower
values are found in the North Polar basin (Ryther, 1963). At present, there
is no adequate information to show that the North Atlantic, where the
ascending water contains little phosphate and other nutrients is relatively
low in primary productivity in comparison with, for example, the North
Pacific. Perhaps the effect of low phosphate concentration is compensated
for by the greater depth and magnitude of ocean currents in the former area.

References

ANDERSON, G. C. and BANSE, K. (1961). Hydrography and phytoplankton production.
  Contr. No. 251, Dept. of Oceanogr. Washington, 1–19.
BURKHOLDER, P. R. and L. M. (1967). Primary productivity in surface waters of the South
  Pacific Ocean. *Limn. Oceanogr.* **12**, 606–617.

CUSHING, D. H. (1969). Upwelling and fish production. *F.A.O. Fish. Techn. Paper No.* **84**, 1–40.

DEACON, G. E. R. (1937). The hydrology of the southern ocean. *Discovery Rep.* **15**, 1–152.

DUGDALE, R. C. (1967). Nutrient limitation in the sea: dynamics, identification, and significances. *Limn. Oceanogr.* **12**, 685–695.

DUURSMA, E. K. (1961). Dissolved organic carbon, nitrogen and Phosphorus in the sea. *Neth J. Sea Res.* **1**, 1–148.

ERIKSSON, E. (1959). The circulation of some atmospheric constituents in the sea. *Rossby Memorial Vol.*, 147–157.

HENTSCHEL, E. and WATTENBERG, H. (1930). Plankton und Phosphat in der Oberflächenschicht des Südatlantischen Ozeans. *Ann. Hydr. Mar. Met.*, **58**, 273–277.

KETCHUM, B. H. (1947). The biochemical relations between marine organisms and their environment. *Ecol. Monogr.* **17**, 309–315.

MENZEL, D. W. (1967). Particulate organic carbon in the deep sea. *Deep Sea Res.* **14**, 229–238.

MONTGOMERY, R. B. (1959). Salinity and residence time of subtropical oceanic surface water. The Atm. and the Sea in motion. Rockefeller and Oxford Un. Press, 143–146.

MUNK, W. H. (1966). Abyssal recipes. *Deep Sea Res.* **13**, 707–730.

POSTMA, H. (1958). Chemical results and a survey of water masses and currents. *Snellius Exp.* 1929–1930, 116 pp.

POSTMA, H. (1964). The exchange of oxygen and carbon dioxide between the ocean and the atmosphere. *Neth. J. Sea Res.* **2**, 258–283.

REDFIELD, A. C., KETCHUM, B. H. and RICHARDS, F. A. (1963). The influence of organisms on the composition of sea-water. *The Sea* **2**, Interscience Publ., 26–77.

REID, J. L. (1962). On circulation, phosphate-phosphorus content, and zooplankton volumes in the upper part of the Pacific Ocean. *Limn. Oceanogr.* **7**, 287–306,

RILEY, G. A. (1951). Oxygen, phosphate and nitrate in the Atlantic Ocean. *Bull. Bingham Oceanogr. Coll.* **12**, 1–126.

RYTHER, J. H. (1963). Geographic variations in productivity. *The Sea* **2**, Interscience Publ., 347–380.

STEELE, J. H. (1957). A comparison of plant production estimates using 14C and phosphate data. *J. Mar. Biol. Ass.* **36**, 233–241.

SVERDRUP, H. U., JOHNSON, M. W., and FLEMING, R. H. (1946). *The Oceans*, Prentice Hall, 1087 pp.

WOOSTER, W. S. and REID, J. L. (1963). Eastern boundary currents. *The Sea* **2**, Interscience Publ., 253–280.

WÜST, G. (1936). *Oberflächensalzgehalt, Verdunstung und Niederschlag auf dem Weltmeere.* Festschr. N. Krebs, Stuttgart, 347–359.

WYRTKI, K. (1967). Water masses in the oceans and adjacent seas, *Int. Dictionary of Geoph.*, Pergamon Press, 1–11.

Terrigenous organic matter
and coastal phytoplankton fertility

A. PRAKASH

Fisheries Research Board of Canada
Marine Ecology Laboratory
Bedford Institute
Dartmouth, N.S., Canada

Abstract

From the point of view of phytoplankton productivity, coastal and oceanic waters are looked upon as two chemically distinct environments, with different productive potentials and possibly different production regimes. The comparatively large amounts of organic matter introduced in the coastal environment through land drainage results in biological conditioning of coastal waters in favour of high phytoplankton production. Particular attention is given to the role which humic compounds play in such biological conditioning.

Resumen

Do ponto de vista da produtividade primária, as águas costeiras e oceânicas são consideradas como dois ambientes quìmicamente distintos, com diferentes potenciais produtivos e, possivelmente, regimes de produção diferentes. A quantidade relativamente grande de matéria orgânica introduzida no ambiente costeiro, através da drenagem, resulta no condicionamento biológico das águas costeiras que favorece a alta produção do fitaplâncton. Particular atenção é dada ao papel que compostos húmicos desempenham nêsse condicionamento biológico.

INTRODUCTION

Any biological appraisal of the actual and potential fertility of the sea must recognize that the coastal waters and the open ocean represent two distinct environments which are different not only in their chemical make up, but also differ considerably in their productive capacities. That the fertility of coastal waters with respect to phytoplankton production is several fold higher as compared with that of oceanic waters appears to be a fair generalization and is equally valid for temperate, sub-tropical and tropical waters, (Anderson, 1964; Menzel, 1959; Teixeira, 1963). Attempts to explain such anomalies in production characteristics of the two ecosystems have revolved around either the "land mass" or "island mass" effect concept (Doty and Oguri, 1956; Jones, 1962; Strickland, 1965), or have been based on enrichment of surface layers brought about by processes like vertical mixing and upwelling (Cooper, 1952).

The greater phytoplankton production in coastal waters* in contrast with that in the open ocean is generally attributed to a higher standing crop in the former which is related in some way to increased supply of inorganic nutrients (mainly N, P and Si) in the euphotic zone. However, there are several situations where high productivity of coastal waters is not explainable entirely on the basis of increased inorganic nutrients (Currie, 1958; Mc-Allister *et al.*, 1960; Provasoli, 1963). While high nutrient levels are suggestive of increased productivity, there appears to be no clear-cut relationship between the degree of fertility and the inorganic nutrient concentration. Thus it is becoming increasingly clear that water masses of identical salinity, temperature and nutrient characteristics may have different biological production potentials and that the potential fertility of a body of sea water may be linked more to the utilization of nutrients by phytoplankton than to its actual nutrient level. The fertility of sea water is now being looked upon in terms of organic as well as inorganic enrichment and the processes of phytoplankton succession and production are explained on the basis of "biological conditioning" which implies presence or absence of biologically active substances in the environment. The ecological significance of biologically active substances viz. vitamins, antibiotics, auxins and other extracellular products is now fairly well recognized (Lucas, 1947; Droop, 1957;

---

* Ryther (1969) has estimated that the primary organic production in the coastal zone which represents less than ten percent of the total area of the world ocean is twice as much as that of the open ocean on an unit area basis.

Provasoli, 1958) and is discussed at length by Provasoli (1963) in his excellent review of organic regulation of phytoplankton fertility.

There are no serious doubts that the input of organic metabolites either in the form of planktonic, phytobenthic and bacterial decomposition products and exudates (autochthonous) or as organic compounds of terrigenous origin (allochthonous) may influence the biological conditioning of sea water for or against the growth of phytoplankton. Thus it can be argued that the differential fertility of coastal and oceanic waters may be linked with the degree of biological conditioning and the so called 'neritic' and 'oceanic' quality of water may, among other factors be a function of relative inputs of autochthonous and allochthonous organic matter. In contrast with the open ocean, allochthonous organic contribution to the coastal environment is indeed impressive, particularly in areas of river discharge, and may account for much of the fertility in inshore coastal waters. Recently Seki *et al.* (1969) have reported that the amount of organic material contributed per year as drainage from land to the Strait of Georgia was comparable to the annual primary production of that area. Terrigenous organic matter washed from the land and that carried to the sea by rivers includes an impressive array of biologically active compounds which directly or indirectly influence the production of phytoplankton in coastal waters. The nature of some of these compounds is reviewed by a number of authors (e.g. Vallentyne, 1957; Fogg, 1959; Koyama, 1962; Duursma, 1965).

A significant fraction of the terrigenous organic matter is represented by humus or humus-like substances and it is often stated that the nearshore coastal waters are particularly rich in such substances (Waksman, 1936; Skopintsev, 1950; Bordovskiy, 1965a; Kalle, 1966). The terrigenous humic material includes, among other things, biochemically stable soil and marsh lechates, plant decomposition products, mangrove exudates etc. Very little is however known of its role in conditioning the inshore waters for overall phytoplankton growth. Humic substances are efficient natural trace metal chelators and it is possible that Gran (1931) recognized this property while suggesting that iron containing humus compounds drained from land may be important for high productivity of coastal waters. Evidence in support of chelation as an important aspect of sea water fertility is mounting and it has been shown that the supply of chelating substances is of paramount importance in biological conditioning of sea water for phytoplankton growth (Johnston, 1964; Barber and Ryther, 1969). Trace element concentrations in the coastal waters are several times higher than in oceanic waters and in view of the large amounts of humic material introduced in inshore coastal environment as a result of land drainage, it would appear that some of these trace elements become available to phytoplankton organisms in more acceptable form and may result in increased primary production. Recent studies (Prakash and Rashid, 1968; 1969) suggest that humic compounds exert a stimulatory effect on the growth of marine phytoplankton and should be looked upon as a significant ecological entity influencing coastal phyto-plankton productivity.

ORIGIN, NATURE AND FATE OF HUMIC MATTER IN COASTAL WATERS

While it is generally believed that humic substances are formed during the decomposition of biogenic material with subsequent polymerisation and polycondensation of phenolic units, there appears to be no unanimity of opinions concerning their exact mode of formation. Opinions of what compounds might constitute the precursors of humic substances have ranged from carbohydrates and proteins to lignins. Regardless of their mode of formation, humic substances can not be looked upon as definite chemical entities but as a dynamic group of heterogenous compounds which are complex in their structural make up and change constantly with time and space. The characterization of the nature, structure and origin of the various groups of compounds taking part in the formation of strictly humic substances is insufficiently known to permit comments on their chemical nature. The separation of humic substances into traditional fractions, e.g. humin, humic, fulvic and hymatomelonic acids (Kononova, 1966) is more an analytical classification than a genetic characterization. There is in-creasing evidence that these are related compounds capable of physico-chemical conversion from one form to another (Thiele and Kettner, 1953).

Humus present in coastal waters largely as a result of drainage from land occurs in dissolved and colloidal state as well as in the form of organic detritus. Most of this humus finds its way to bottom sediments through adsorption on clay or other mineral particles as stable insoluble organo-mineral compounds (Bordovskiy, 1965b), while the remainder stays in inshore surface waters and is available to plankton and filter-feeding or-ganisms (Fig. 1). The characteristic yellowish-brown colour of near shore coastal water is due mainly to dissolved humus of terrigenous origin and is represented largely by low molecular weight fulvic and humic acids. Other autochthonous substances possessing the properties of fulvic and humic acids, viz. planktonic and phytobenthic decomposition products and exudates, are also a source of yellow colour in sea water and have been referred to as 'Gelbstoff' (Kalle, 1966) or 'Water humus' (Skopintsev, 1950). The water soluble yellow fulvo-humic fraction shows fluorescence in UV light and this property has been used quite effectively in measuring 'Gelbstoff' or marine humus in sea water (Kalle, 1949). The direct relationship

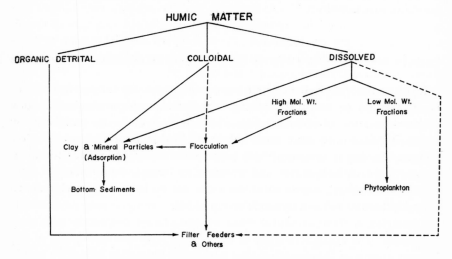

Figure 1   Fate of terrigenous humic matter in inshore waters

between yellow colour of the sea water and salinity as demonstrated by Kalle (1949) and Jerlov (1955) suggests that for the most part, the yellow UV fluorescing substance in inshore coastal waters is contributed by the fresh water run off from land. This is generally true of a number of embayments on the Atlantic coast of Canada.

Commenting on the fate of Gelbstoff in sea water, Sieburth and Jensen (1968) maintain that humic materials of terrestrial origin precipitate soon after coming in contact with sea water and as such have a very short residence time in solution. However this has not been borne out by our observations in the coastal waters of New Brunswick and Nova Scotia which receive substantial amounts of humus-laden water from land particularly during the spring run off. According to our observations, a limited amount of flocculation or precipitation of humic material occurs in the sea but it is a slow and incomplete process and involves mainly the high molecular weight humic fractions. The specific gravity of these flocculates is similar to that of sea water and it seems unlikely that they are incorporated in bottom sediments without the help of clay or other suspended mineral particles. However the possibility that they are very similar to organic aggregates of Riley (1963) and may be utilized by filter feeding organisms as suggested by Sieburth and Jensen (1968) appears attractive. Direct utilization of dissolved humic matter by marine animals may appear unlikely, but in view of the presence of amino acids in the structure of humic compounds (Bremner, 1955) and the capacity of certain marine animals to remove amino acids from sea water (Stephens and Schinske, 1961), this possibility cannot be completely ruled out.

Most of the dissolved humus in river water and coastal inshore waters is represented by relatively low molecular weight fulvo-humic fractions which show no signs of flocculation in the presence of sea water. Molecular weight distribution of humic acids as determined by Sephadex gel-filtration method (Rashid and King, 1969) suggests that the humic acids found in dissolved state in yellow river water, mangrove swamps and inshore coastal waters are mainly composed of fractions having molecular weights less than 1500 and rarely exceeding 5000 (Table 1).

INFLUENCE OF HUMIC SUBSTANCES ON PHYTOPLANKTON GROWTH

While limnologists and plant physiologists have been aware of the ecological importance of water humus in the growth of aquatic and terrestrial plants for over thirty years, surprisingly little is known of the influence of humic substances on marine algae. Although suggestions have been made that humic substances may stimulate phytoplankton production in the sea (Provasoli, 1963; Strickland, 1965; Droop, 1966), there has been very little direct evidence to support them. Current thoughts about the positive effect of humic substances on marine phytoplankton growth appear to be largely influenced by the successes which have been achieved in cultivating a number of phytoplankton species in media enriched with aqueous extracts of soil. The growth promoting property of soil extract has been attributed mainly to the chelating action of its humic component and to a lesser extent to its contribution of vitamins and other accessory growth factors.

Table 1 Molecular weight distribution of humic acids extracted from various sources (modified after Prakash and Rashid, 1968)

Sephadex Fraction	Mol. Wt range	Soil extract (aqueous) %	Mangrove lechate %	Digdeguash River, New Brunswick			Forest soil %	Marine sediment %
				Main River %	River mouth (S‰=25.31) %	Passam-aquoddy Bay (S‰=28.94) %		
I	<700	43.8	86.4	63.0	71.1	68.3	33.2	6.5
II	700– 1,500	56.2	13.6	26.8	20.5	23.9	7.7	5.0
III	1,500– 5,000	—	—	11.2	8.4	7.8	11.9	5.6
IV	5,000–10,000	—	—	—	—	—	47.2	9.9
V	>10,000	—	—	—	—	—	—	73.0

Direct observations on the effect of humic substances on unialgal cultures of phytoplankton have been few and somewhat casual, nevertheless they have provided useful information. Wilson and Collier (1955) noticed an improvement in growth of *Gymnodinium brevis* when water from a Florida river and extract of peat soil were added to the culture medium. Likewise, Droop (1966) observed that the growth of *Skeletonema costatum* in a defined medium was significantly increased when humic fraction of soil humus was added to the medium. Recently Prakash and Rashid (1968) used purified fractions of humic acid, fulvic acid and hymatomelonic acid extracted from yellow river water, soil and marine sediments for their enrichment experiments and found some of them to be stimulatory for dinoflagellates. They showed that the addition of humic and fulvic acids in small amounts to the culture media enhanced the growth rate and yield of marine dinoflagellates and that the growth responses were generally dependent on concentration. For the same concentration, humic acid gave a higher growth response than fulvic acid, whereas, hymatomelonic acid remained ineffective (Fig. 2 and Table 2).

One of the consistent features in all our humic enrichment experiments has been the high growth response of diatoms and dinoflagellates to low mole-

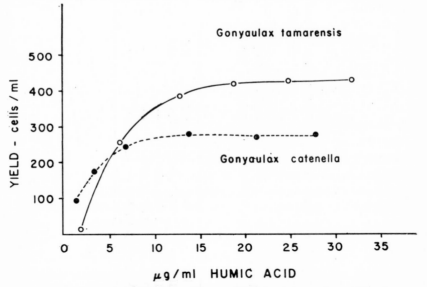

Figure 2 Response of dinoflagellates to varying concentrations of humic acid (after Prakash and Rashid, 1968)

Table 2 Effect of humic, fulvic and hymatomelonic acids on the growth of the dino-flagellate *Gonylax tamarensis* (Prakash and Rashid, 1968).

	Concentration (μg/ml)	Final yield (cells/ml)	Growth Constant K_{10} (days)$^{-1}$	Generation time (tg) (hr)
Humic acid	14.1	1179	0.10	72.4
	7.0	1086	0.10	72.6
	3.5	837	0.08	90.3
Fulvic acid	7.5	602	0.07	103.2
	3.7	451	0.05	144.5
	0.9	283	0.02	361.2
Hymatomelonic acid	10.5	no appreciable growth		
	5.3	no appreciable growth		

cular weight fractions of humic acid. Sephadex fraction I (mol. wt. <700) was generally the most effective and generated higher growth response in cultures relative to other fractions (Figs. 3 and 4). Since such low molecular weight fractions are predominant in inshore coastal waters (Table 1), there seems little doubt that they influence the biological conditioning of sea water in favour of increased phytoplankton growth.

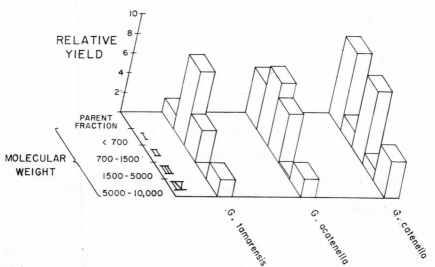

Figure 3 Growth response of *Gonyaulax* to different molecular weight fractions of humic acid. Each fraction was added at 4 μg/ml concentration. Yield refers to maximum cell numbers obtained with different fractions

The positive effect of humus substances is not only reflected in increased growth rate and cell yield, but also in the rate of C^{14} assimilation by phytoplankton. Cultures grown in medium enriched with humic acid show a

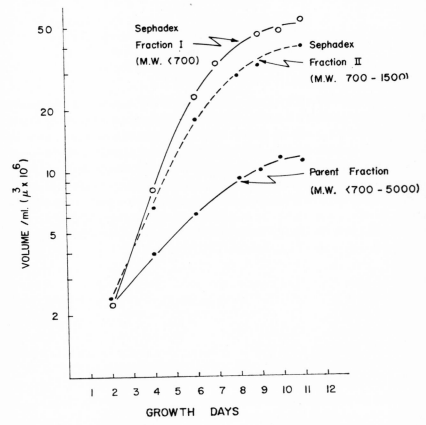

Figure 4 Growth response of *Thalassiosira nordenskioldii* to the various fractions of humic acid. Volume refers to total cell volume of a culture as determined by means of a Coulter counter and an automatic particle size distribution analyzer

marked increase in C^{14} uptake compared with those grown in medium devoid of humic acid (Table 3). Furthermore, the rate of C^{14} assimilation is not only dependent on the concentration of the humic additive, but is also influenced by humic fractions of different molecular weights (Fig. 5).

Table 3 Influence of humic acid on C^{14} assimilation by marine dinoflagellates

Species	Incubation period	Medium without humic acid	Medium with umic acid
	hr	cpm	cpm
Gonyaulax tamarensis	3	3850	5260
Gonyaulax catenella	3	4350	5600
Gonyaulax monilata	2	1250	1710

Prakash and Rashid (1968) found that the rate of C^{14} assimilation by marine dinoflagellates generally increased with increasing concentration of humic acid in the culture medium, but at high concentrations of humic acid the radiocarbon assimilation was significantly reduced. Similar responses have

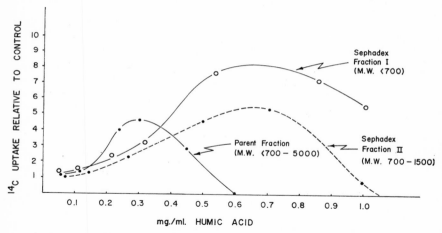

Figure 5 Influence of humic enrichment on C^{14} assimilation by *Thalassiosira nordenskioldii*

also been observed in diatom cultures. However, the reasons for inhibition of C^{14} uptake at higher concentrations of humic acid are not clearly understood.

So far the evidence presented for growth stimulating effect of humic compounds has been based on experiments carried out on laboratory cultures of phytoplankton. There are however, certain observations made on natural phytoplankton populations which support the results obtained with culture experiments and are worth mentioning here. In 1968, I had the opportunity of carrying out an investigation on the influence of mangrove exudates on dinoflagellate blooms in Oyster Bay which is located near Montego Bay in Jamaica, West Indies. This Bay is famous for its bio-luminescence cause by a year-round persistent bloom of *Pyrodinium baha-mense* and has been the object of an intensive study by a group from the John Hopkins University (Taylor *et al.*, 1966; Carpenter and Seliger, 1968). It was observed that the growth of *Pyrodinium bahamense* was substantially improved in the presence of yellowish-brown run off from mangrove swamps bordering the Bay. The generation time of this dinoflagellate within the Bay was estimated to the between 60–63 hr; however, towards the mangrove swamps, the growth conditions appeared to be good and an average genera-tion time of 59 hr was calculated. Observations made on *Pyrodinium* isolated from Oyster Bay and grown in natural Bay water as well as in a marine synthetic medium (ESWA)* revealed that the growth was stimulated by the addition of mangrove lechate in small amounts to the culture medium (Table 4). Addition of various vitamins to culture media did not improve

Table 4 Growth *Pyrodinium bahamense* in media with and without humic enrichment

Medium	Humic acid (μg/ml)	Growth Constant K_{10} (days)$^{-1}$	Generation Time tg (hr)
Oyster Bay water	—	0.114	63.4
Oyster Bay water + Mangrove lechate	7.5	0.124	59.0
Marine Synthetic Medium (ESWA)	—	0.100	72.2
ESWA + Mangrove lechate	7.5	0.120	60.2

the growth but the addition of trace metals along with the yellowish-brown water decidedly enhanced the growth of *Pyrodinium*. Analysis of the man-grove lechates indicated that their humic component was represented almost exclusively by Sephadex fractions I and II (Table 1). These fractions ap-parently provided a continuous source of chelating and sensitizing agents which kept the dinoflagellates in an exponential state of growth. This beneficial action of humic additives which enables a phytoplankton species to grow to bloom proportions may also explain the development of 'red tides' in coastal environment. Sufficient indirect evidence has accumulated which suggests that heavy rainfall or land drainage is a prerequisite to most,

* for composition of this medium refer to Prakash and Rashid (1968)

if not all, dinoflagellate blooms in coastal waters and that the intensity of the bloom may be related to humic or other terrigenous nutritional factors entering the inshore waters (Slobodkin, 1953; Nümann, 1957; Steven, 1966; Burkholder *et al.*, 1967).

PHYSIOLOGICAL EFFECTS INDUCED BY HUMIC SUBSTANCES

This section is largely speculative, since no specific study of the effect of humic substances on the physiology of marine unicellular algae has been reported to the best of my knowledge. On the other hand, considerable data on the influence of humic compounds on the metabolic processes of higher plants are available (Guminski, 1969) which through inference or analogy may be used to elucidate processes involved in growth stimulation of marine phytoplankton under humic enrichment conditions.

One of the main functions of humic substances is of course chelation, which plays a vital role in nutritional physiology of higher plants as well as of unicellular algae. As efficient natural chelating agents, humic compounds not only control the supply of trace metals to the plant cell, but may also influence the ionic balance within the cell. The growth stimulating property of humic and fulvic acids has been attributed chiefly to their ability to peptise and hold sufficient amounts of iron and other trace elements in solution for eventual assimilation by the plant cell (Burk *et al.*, 1932; De Kock, 1955; D'Yakonova, 1962; Shapiro, 1966). The amount of ionic iron in the sea is extremely small for effective utilization by the phytoplankton and most of the iron present in sea water occurs as insoluble ferric hydroxide which is unavailable to planktonic algae unless adequately chelated. It appears that metal-organic complexes formed by humic compounds and resulting in growth stimulation are of Fe-humate and Fe-fulvate type and provide iron to the algal cell in more acceptable form. There is evidence to suggest that such metal-organic complexes are capable of entering the plant cell and that the degree of penetration depends on their molecular size (Prát *et al.*, 1961; Aso and Sakai, 1963). Thus, it is possible to account for the higher growth response noticed in algal cultures when enriched with low molecular weight fractions of humic acid (Figs. 3 and 4). The case for the control and supply of trace metals to unicellular algae through chelation, in the light of recent evidence, is a strong one. However, a generalized theory explaining the physiological stimulation and pathways leading to growth enhancement has not yet emerged. Theoretical aspects of chelation

4*

chemistry with respect to iron and its ecological implications are discussed by Droop (1961) and by Johnston (1964) who concluded that "the solubilization of trace metals is a most important aspect of sea water fertility in the presence of adequate nutrients".

Aside from the chelation aspect, humic compounds have been shown to affect the growth and yield of higher plants by stimulating the physiological and biochemical processes associated with cellular metabolism. Studies by several authors and particularly those of Prozorovskaya (1936), Khristeva (1953) and Chaminade (1956) suggest that the humic substances act as specific sensitizing agents and enhance the permeability of the plant cell membrane thereby increasing the uptake of nutrients from the surrounding medium. When used in small amounts, humic compounds have been shown to increase activity of a number of enzymes, e.g. invertase (Vaughan, 1967), glutamylalanine transaminase (Cincerova, 1964), acid and alkaline phosphatases (Scheffer *et al.*, 1962) etc. Recent experimental evidence indicates that humic substances may play a significant role in intensifying protein, DNA and RNA metabolism in higher plants (Khristeva *et al.*, 1967; Fialova, 1969). Humic substances have also been shown to act as respiratory catalysts since oxygen is absorbed more intensively by the plant cells in their presence (Biber and Magaziner, 1951; Flaig, 1958; Smidova, 1960). This increase in plant cell respiration is explained by Khristeva (1953) as due to the presence in humic compounds of quinone groups which are not only hydrogen acceptors but also act as oxygen activators, furthermore, humic acids entering the plant cells become a supplementary source of polyphenols which function as respiratory catalysts.

This has been not more than a mere glimpse of the many sided physiological effects induced by humic compounds on plant cells and it is beyond the scope of this presentation to review the various mechanisms and physiological pathways involved in altering metabolic processes. The positive effect of humic compounds on the growth of higher plants has undoubtedly been established and if one assumes that the metabolism of planktonic algae is fundamentally similar to that of higher plants as implied by Arnon (1958), then many of the observations made on the growth stimulation of higher plants may be equally valid for marine phytoplankton. The identification and isolation of biologically active components of humic matter that stimulate growth of marine phytoplankton must await further investigations, nonetheless the ecological influence of such compounds on phytoplankton fertility is obvious.

Acknowledgement

I am indebted to Dr. M. A. Rashid of Marine Geochemistry Group, AOL, Bedford Institute, for many helpful discussions and suggestions during the development of this paper. His valuable help in the various chemical analyses is also gratefully acknowledged.

References

ANDERSON, G. C. (1964). The seasonal and geographical distribution of primary productivity off the Washington and Oregon Coasts. *Limnol. Oceanogr.* **9**, 284–302.

ARNON, D. I. (1958). Some functional aspects of inorganic micronutrients in the metabolism of green plants. In *Perspectives in Marine Biology* pp. 351–383 (ed. Buzzati-Traverso, A. A.). Univ. California Press.

ASO, S. and SAKAI, I. (1963). Studies on the physiological effects of humic acid. I. Uptake of humic acid by crop plants and its physiological effects. *Soil Sci. and Plant Nutr.*, Tokyo. **9**, 3, 1.

BARBER, R. T. and RYTHER, J. H. (1969). Organic chelators: Factors affecting primary production in the Cromwell current upwelling. *J. Exp. Mar. Biol. Ecol.*, **3**, 191–199.

BIBER, V. and MAGAZINER, K. (1951). The effect of humic and fulvic acids on the respiration of isolated plant tissues. *Dokl. Akad. Nauk. SSSR.* **76**, 4, 609.

BORDOVSKIY, O. K. (1965a). Accumulation and transformation of organic substances in marine sediments. 2. Sources of organic matter in marine basins. *Mar. Geol.*, Special issue. **3**, 5–31.

BORDOVSKIY, O. K. (1965b). Accumulation and transformation of organic substances in marine sediments. 3. Accumulation of organic matter in bottom sediments. *Mar. Geol.*, Special issue **3**, 33–82.

BREMNER, J. M. (1955). Studies on soil humic acids. I. The chemical nature of humic nitrogen. *J. Agr. Sci.* **46**, 247–256.

BURK, D., LINEWEAVER, H. and HORNER, C. K. (1932). Iron in relation to the stimulation of growth by humic acid. *Soil Sci.* **33**, 413–453.

BURKHOLDER, P. R., BURKHOLDER, L. M., and ALMADOVER, R. L. (1967). Carbon assimilation of marine flagellate blooms in neritic waters of southern Puerto Rico. *Bull. Mar. Sci. Gulf. Caribbean.* **17**, 1–15.

CARPENTER, J. H. and SELIGER, H. H. (1968). Studies on a bioluminescent Bay. II. Flow patterns and exchange rates. *J. Mar. Res.* **26**, 3, 256–272.

CHAMINADE, R. (1956). Action de l'acide humique sur le développement et la nutrition minérale des végétaux. *Trans. 6th Intern. Congr. Soil Sci.* Paris. **4**, 65.

CINCEROVA, A. (1964). Effect of humic acid on transamination in winter wheat plants. *Biol. Plant.* No. **6**.

COOPER, L. H. N. (1952). Processes of enrichment of surface water with nutrients due to strong winds blowing on to continental slope. *J. Mar. Biol. Assoc. U. K.* **30**, 3, 453–464.

CURRIE, R. I. (1958). Some observations on organic production in the northeast Atlantic. *Rapp. Cons. Explor. Mer.* **144**, 96–102.

DE KOCK, P. C. (1955). Influence of humic acids on plant growth. *Science* **121**, 473–474.

DOTY, M. S. and OGURI, M. (1956). The island mass effect. *J. Cons. Int. Explor. Mer.* **22**, 1, 33–37.

DROOP, M. R. (1957). Auxotrophy and organic compounds in the nutrition of marine phytoplankton. *J. Gen. Microbiol.* **16**, 286–293.

DROOP, M. R. (1961). Some chemical considerations in the design of synthetic culture media for marine algae. *Botanica Marina.* **2**, 3/4, 231–246.

DROOP, M. R. (1966). (Comments. Proc. 2nd International Interdisciplinary Conference). In *Marine Biology* II. pp. 158–159 (ed. Oppenheimer, Carl H.). N. Y. Acad. Sci. New York.

DUURSMA, E. K. (1965). The dissolved organic constituents of sea water. In *Chemical Oceanography.* pp. 432–475 (ed. Riley, J. P. and Skirrow, G.). Academic Press.

D'YAKONOVA, K. V. (1962). Iron-humus complexes and their role in plant nutrition. *Soviet Soil Sci.* No. 7, 692–698.

FIALOVA, S. (1969). Influence of sodium humate and nutritive conditions on the content of nucleic acid, particularly the Ribosomal Ribonucleic acid in wheat roots. *Biol. Plant.* **11**, 1, 8–22.

FLAIG, W. (1958). Die Chemie organischer Stoffe im Boden und deren physiologische Wirkung. Verhandlungen d. II. und IV. *Kommission d. Internat. Bodenkundlichen Gesellschaft.* Hamburg. **2**, 11–45.

FOGG, G. E. (1959). Dissolved Organic matter in oceans and lakes. *New Biology.* **29**, 31–48.

GRAN, H. (1931). On the conditions for the production of plankton in the sea. *Rapp. Cons. Int. Explor. Mer.* **75**, 37–46.

GUMIŃSKI, S. (1969). Present day views on physiological effects induced in plant organisms by humic compounds. *Soviet Soil. Sci.* No. 9, 1250–1256.

JERLOV, N. G. (1955). Factors influencing the transparency of the Baltic waters. *K. Vet. O. Vitterh. Samh. Handl.* F6. Ser. B. Bd. 6. No. **14**.

JOHNSTON, R. (1964). Sea water, the natural medium of phytoplankton. II. Trace metals and chelation, and general discussion. *J. Mar. Biol. Assoc. U. K.*, **44**, 87–109.

JONES, E. C. (1962). Evidence of an island effect upon the standing crop of zooplankton near the Marquesas Islands, Central Pacific. *J. Cons. Int. Explor. Mer.* **27**, 3, 223–231.

KALLE, K. (1949). Fluoreszenz und Gelbstoff im Bottnischen und Finnischen Meerbusen. *Deutsche Hydrog. Z.* **2**, 117–124.

KALLE, K. (1966). The problem of Gelbstoff in the sea. In *Oceanography and Marine Biology. Annual Review* **4**, pp. 91–104 (ed. Barnes, H.). George Allen & Unwin Ltd. London.

KHRISTEVA, L. A. (1953). The participation of humic acids and other organic substances in the nutrition of higher plants. *Pochvovedenie.* **10**, 46–59.

KHRISTEVA, L. A., SOLAKHA, K., DINKINA, R., KOVALENKO, V., and GOROVAYA, A. (1967). Influence of physiologically active substances in soil humus and fertilizers on nucleic acid conversion, plant growth, and their after effect on seed qualities in generations. Symposium "*Humus et Planta*" IV. Praha.

KONONOVA, M. M. (1966). *Soil organic matter, its nature, its role in soil formation and in soil fertility.* 2nd Ed. Pergamon Press Ltd. 544 p.

KOYAMA, T. (1962). Organic compounds in sea water. *J. Oceanogr. Soc. Japan.*, 20th Anniv. Vol., 563–576.

LUCAS, L. E. (1947). The ecological effects of external metabolites. *Biol. Rev.* **20**, 270–295.

MCALLISTER, C. D., PARSONS, T. R. and STRICKLAND, J. D. H. (1960). Primary production at "station P" in the Northeast Pacific Ocean. *J. Cons. Int. Explor. Mer.* **25**, 3, 240–259.

MENZEL, D. W. (1959). The „Island mass effect" at Bermuda. In *The plankton ecology and related chemistry and hydrography of the Sargasso Sea.* Prog. Rept. Sept. 1967 to April 1959. AEC Contract No. AT(30-1)-2078) (unpublished).

NÜMANN, W. (1957). Natürliche und Künstliche "red water" mit anschließendem Fischsterben im Meer. *Arch. Fischereiwiss.* **8**, 3, 204–209.

PRAKASH, A. and RASHID, M. A. (1968). Influence of humic substances on the growth of marine phytoplankton: Dinoflagellates. *Limnol. Oceanogr.* **13**, 4, 598–606.

PRAKASH, A. and RASHID, M. A. (1969). The influence of humic substances on coastal phytoplankton productivity. In *Coastal Lagoons—a Symposium.* UNESCO and UNAM. Mexico City. Nov. 1967. (In press).

PRAT, S., SMIDOVA, M. and CINCEROVA, A. L. (1961). Penetration and effect of humus substances (fractions) on plant cells. *Fifth Intern. Congr. Biochem. Abstracts of Commun.*, **329**. Moscow 1961.

PROVASOLI, L. (1958). Growth factors in unicellular marine algae. In *Perspectives in Marine Biology.* pp. 385–403 (ed. Buzzati-Traverso, A.A.) Univ. California Press.

PROVASOLI, L. (1963). Organic regulation of phytoplankton fertility. In *The Sea.* Vol. **2**, pp. 165–219 (ed. Hill, M. N.). Interscience. London.

PROZOROVSKAYA, A. A. (1936). The effect of humic acid and its derivatives on the uptake of nitrogen, phosphorus, potassium and iron by plants. Collected papers. *Organomineral Fertilizers* (Organomineral'nye udobreniya). Trudy. nach.-issles. Inst. udobr. insektisid. Fungisid. **127**.

RASHID, M. A. and KING, L. H. (1969). Molecular weight distribution measurements of humic and fulvic acid fractions from marine clays on the Scotian shelf. *Geochim. Cosmochim. Acta.* **33**, 147–151.

RILEY, G. A. (1963). Organic aggregates in sea water and the dynamics of their formation and utilization. *Limnol. Oceanogr.* **8**, 372–381.

RYTHER, J. H. (1969). Photosynthesis and fish production in the sea. *Science.* **166** (3901), 72–76.

SCHEFFER, F., ZIECHMANN, V. and ROCHUS, V. (1962). Die Beeinflussung von Phosphatase-Aktivitäten durch Huminstoffe. *Naturwiss.* **19**.

SEKI, H., STEPHENS, K. V. and PARSONS, T. R. (1969). The contribution of allochthonous bacteria and organic materials from a small river into a semi-enclosed area. *Arch. Hydrobiol.* **66**, 1, 37–47.

SHAPIRO, J. (1966). The relation of humic color to iron in natural waters. *Verh. Intern. Verein. Limnol.* **16**, 477–484.

SIEBURTH, J. McN. and JENSEN, A. (1968). Production and transformation of extracellular organic matter from littoral marine algae; a resume. Symposium. *Organic matter in natural waters.* College. Alaska. Sept. 1968 (in press).

SKOPINTSEV, B. A. (1950). Organic matter in natural waters (water humus). *Trudy Geol. Inst. Akad. Nauk. SSSR.* **17** (29).

SLOBODKIN, L. B. (1953). A possible initial condition for red tides on the coast of Florida. *J. Mar. Res.* **12**, 148–155.

SMIDOVA, M. (1960). The influence of humus acid on the respiration of plant roots. *Biol. Plant.* **2**, 152–164.

STEPHENS, G. C. and SCHINSKE, R. A. (1961). Uptake of amino acids by marine invertebrates. *Limnol. Oceanogr.* **6**, 2, 175–181.

STEVEN, D. M. (1966). Characteristics of a red water bloom in Kingston Harbour, Jamaica. W. I. *J. Mar. Res.* **24**, 2, 113–123.

STRICKLAND, J. D. H. (1965). Production of organic matter in the primary stages of the marine food chain. In *Chemical Oceanography.* pp. 477–610 (ed. Riley, J. P. and Skirrow, G.). Academic Press. London.

TAYLOR, W. R., SELIGER, H. H., FASTIE, W. G. and McELROY, W. D. (1966). Biological and physical observations on a phosphorescent Bay in Falmouth Harbour, Jamaica. W. I. *J. Mar. Res.* **24**, 1, 28–43.

TEIXEIRA, C. (1963). Relative rates of photosynthesis and standing stock of net phytoplankton and nannoplankton. *Bol. Inst. Oceanogr. Universidade de Sao Paulo* **13**, 53–60.

THIELE, H. and KETTNER, H. (1953). Über Huminsäuren. *Kolloid. Z.* **130**, 131–160.

VALLENTYNE, J. R. (1957). The molecular nature of organic matter in lakes and oceans, with lesser reference to sewage and terrestrial soils. *J. Fish. Res. Bd. Canada* **14**, 1, 33–82.

VAUGHAN, D. (1967). Effect of humic acid on the development of invertase activity in slices of beetroot tissue washed under aseptic conditions. Symposium "*Humus et Planta*" IV. Praha.

WAKSMAN, S. A. (1936). *Humus, origin, chemical composition and importance in nature.* Baillière, Tindall and Cox. London.

WILSON, W. B., and COLLIER, A. (1955). Preliminary notes on the culturing of *Gymnodinium brevis*, Davis. *Science* **121**, 394–395.

Nutritional relationships in marine organisms

L. PROVASOLI

Haskins Laboratories
165 Prospect Street
New Haven, Connecticut 06520

INTRODUCTION

Lucas in 1947 developed his broad concept of non-predatory relationships linking marine organisms. Metabolites (ectocrines) produced by one kind of organism may affect others by producing necessary nutrients, inhibitory compounds, or by supplying hormones.

His hypothesis was documented by the finding of Léfèvre *et al.* (1952) that some freshwater algae produce auto- and hetero-antagonists; that many marine and freshwater algae require vitamins (Provasoli and Pintner, 1953); that these vitamins are present in waters and produced by microbes and some algae (Robbins *et al.*, 1952); that antibiotics are produced by algae and other marine organisms (Prat *et al.*, 1951; the Burkholders, 1958; Sieburth, 1959); and that algae excrete various nutrients (Fogg and West-lake, 1955; Lewin, 1956; etc.). Progress in these fields is easily traced from reviews (to name a few: Nigrelli, 1958; Jorgensen, 1962; Provasoli, 1963; Hellebust, 1965; Strickland, 1965; Fogg, 1966; Sieburth, 1968; and Menzel and Ryther, 1969).

Of all environments, water most favors medium-mediated interactions between organisms: the organisms live, secrete, die and decompose in this common medium. Since many biologically-active substances are water soluble, diffusion and water movements effect rapid exchange of substances within the community, favoring consumption and dilution of any utilizable substance. The amount of nutrients and metabolites found in waters being the result of a dynamic process governed by rate of excretion, diffusion and uptake, can be a deceptive index of the events; the same amount may be the balance of a sluggish as well as of an extremely active turnover.

Another difficulty is the problem of extracting microgram or nanogram quantities from a medium containing over 3000 mg per 100 ml of assorted inorganic salts. Some ingenious ways have been devised: chelators are trapped by chromatography on metal-impregnated ligand-rich exchange resins (Siegel, 1967); vitamins are bioassayed by means of algae combining sensitivity and hardiness (Ryther and Guillard, 1962; Gold, 1964; Carlucci and Silbernagel, 1966–67); the biological potential of waters is determined by the effects of differential enrichments and other additions on the ^{14}C uptake and growth of natural populations or selected phytoplankters (Menzel *et al.*, 1963; Johnston, 1964; Smayda, 1969; Thomas, 1969; Barber and Ryther, 1969); the production and secretion of metabolites is followed by chemical and biological analysis of waters or media in which large quantities of organisms have been grown or concentrated and allowed to metabolize (Fogg, 1966; Craigie and McLachlan, 1964; Hellebust, 1965). Each method has particular advantages; each has furthered understanding.

Equally successful has been the approach of using pure cultures of organisms grown in defined media to study nutritional needs and production of factors which might act on other organisms. The exclusion of other organisms, and their effects, and of biologically active substances which might be present in natural waters often reveals deficiencies. These deficiencies can sometimes be overcome by known substances. When this fails, syntrophic growth with other organisms of the same environment may relieve the deficiency symptoms, leading to assay of the effects of supernatants and ultimately to identification of the active principles.

Recent results obtained with this approach suggest the need to reconsider the cycle and ecological relevance of vitamins in waters and to review other relationships among aquatic organisms. Marine and fresh water forms will be considered together since data are few and since there is no evidence that organisms of these two environments differ basically either in their needs for vitamins or in their ability to release extracellular products.

VITAMIN EXCRETION BY ALGAE

Factors triggering blooms and governing temporal succession of phyto-plankton species are receiving increasing attention (Fogg, 1969). In both cases the predominance of one or few species suggests strongly that special organismal idiosyncrasies are met by unique ecological opportunities. Any combination of environmental factors such as temperature, light, total solids and nutrient levels may influence or perhaps determine algal pre-dominance at a particular time and location. However, despite much work, the marine algae seem to differ little nutritionally; most are photoauto-trophic; a few species have some heterotrophic abilities, and there are indi-cations of differences in trace metal requirements. The most conspicuous idiosyncrasy seems to be their species-specific requirement for vitamins.

So long as vitamins had been thought to be mainly bacterial in origin, the specificity of algae for vitamins did not seem enough of a differential to be a significant factor for blooms or succession. Historically the bacterial role in the vitamin cycle in the sea came about during the flurry of work caused by the discovery of vitamin B_{12} and of its need by many micro-organisms. A high incidence of B_{12} producers was found among marine bacteria collected from waters and muds or epiphytic on seaweeds (Ericson and Lewin, 1953; Burkholder and Burkholder, 1956, 1958); these bacteria release B_{12} into the medium.

The role of algae in the vitamin cycle was considered to be mainly that of favoring large populations of decomposing bacteria at the end of each algal pulse (review, Provasoli, 1963). The recent proof that 3 marine algae release vitamins in the culture medium calls for re-appraising the vitamin cycle in the sea. Carlucci (1969, and personal communication) followed the concentrations of B_{12}, thiamine, and biotin during the growth cycle of bacteria free cultures of *Gonyaulax polyhedra, Stephanopyxis turris, Skele-tonema costatum* and *Coccolithus huxleyi*. The algae were grown in media containing only the vitamin required by each of them (at limiting and non-limiting concentrations). *G. polyhedra, S. turris,* and *S. costatum*, which re-quire B_{12}, excreted thiamine and biotin; *C. huxleyi*, a thiamine requirer, excreted biotin and vitamin B_{12}. When the organisms are grown at limiting concentrations of the needed vitamins, excretion of the vitamins which they synthesize is less but still significant, except for the *C. huxleyi* which, at limiting concentrations of thiamine, does not excrete B_{12} while still releasing good quantities of biotin. Vitamin production per cell varies during the growth cycle and is in general high in the log and stationary phases.

Ford and Goulden (1959) reported earlier that *Ochromonas malhamensis,* a freshwater chrysomonad requiring a B_{12}, thiamine and biotin, excretes vitamins of the B_6 group, nicotinic and pantothenic acids. The ratio between vitamin cell content and culture fluid content was 1 : 6 for pantothenic acid, 1 : 3 for nicotinic acid, and 1 : 15 for the vitamin B_6 group, indicating that even during the period of exponential growth loss of vitamins from the organism to the surrounding medium was large and "constituted a burden on synthesis". Unpublished results of S. Aaronson, H. Baker and O. Frank show that *O. danica*, a freshwater biotin and thiamine requirer, excrete at midlog relevant quantities of folic, nicotinic and pantothenic acids, but apparently no vitamin B_{12}; again the cells excrete in the medium from 2 to 7 times their cell content in vitamins. Release of extracellular nicotinic acid by *Chlamydomonas* was reported by Nakamura and Gowans (1964).

Zehnder (1949) found that the blue-green algal partner of some lichens excreted substantial quantities of carbohydrate and thiamine which are needed by the fungal partner; another lichen alga *Coccomyxa* excreted biotin (Bednar and Holm–Hansen, 1964). Burkholder (1963) in a short paragraph lacking details, stated that B vitamins were produced "during growth of pure cultures of green algae, blue green algae and diatoms isolated from marine muds", as evidenced by the halo growth of vitamin-requiring bacteria on a vitamin-free agar medium. Similarly, par is of algae needing different vitamins can be grown in vitamin-free media demonstrating an active exchange of needed metabolites. Droop (1968, p. 715) grew *Mono-chrysis lutheri* (requiring B_{12}) syntrophically with *Nannochloris oculata* which does not require vitamins; A. F. Carlucci (personal communication) grew syntrophically *Coccolithus huxleyi* (requiring thiamine) with *Skeleto-nema costatum* which requires B_{12}.

Algae and seaweed are known to be rich in vitamins. Now we can add that they can *synthesize the vitamins that they do not require* (as do all other organisms) and that excretion of surplus vitamins may be widespread since it occurred in fresh water and marine taxa belonging to several ecologically important algal groups. In fact, since excretion of different metabolites seems common in algae (Fogg, 1962) it would be surprising if vitamin excretion was the only exception.

If so, the algae may well be the major producers of vitamins in the sea, for algal biomass exceeds bacterial biomass, except perhaps in highly polluted locations. As a consequence any substantial algal growth would, by consuming the vitamins needed and by excreting the vitamins synthesized, change substantially and pre-condition the waters. This pre-conditioning might be even more significant ecologically than the well known depletion in N and P and Si caused by previous blooms. The latter change is non-specific and at best would favor oligotrophic species, with more active uptake and smaller size. The change caused by simultaneous depletion of some vitamins and enrichment of others, is governed by the specificity of the predominant species in their vitamin requirements, resulting in a specific excretion of the vitamins which are not required (i.e. synthetized). A bloom of a B_{12} requirer would deplete waters of B_{12} and enrich them with biotin and thiamine (to consider only the vitamins needed by the algae); an auto-troph will excrete the 3 vitamins, etc.

The species-specific requirement for vitamins would then impinge on the next algal growth in determining what vitamins will be available and there-fore favoring the species whose species-specificity matches availability. This hypothesis does not conflict grossly with what is presently known. The emerging trends in vitamin requirements of algal groups (Table 1) seems to fit with the general type of succession found in temperate waters: an early spring diatom bloom when B_{12} and other nutrients are high (predominantly

Table 1   Summary of vitamin requirements

Algal Groups	No Vitamins	Vitamins	B_{12}	Thiamine	$B_{12} +$ Thiamine	Biotin	$B_{12}/$ Thiamine	Predomins Vitamins
1) Do not require biotin								
Blue-green algae	9	7	7	0	0		7/0	B_{12}
Chlorophytes	24	49	16	13	20		36/33	0
Diatoms	22	24	18	3	3		21/6	B_{12}
Cryptomonads	0	11	2	2	7		9/9	0
2) Require biotin								
Euglenids	0	11		1	9	1		0
Chrysomonads	1	27	1	9	14	5	17/24*	Thia.
Dinoflagellates	1	18	13	0	4	5	17/5†	B_{12}

Not in table are: * 2 *Synura* requiring B_{12} + biotin; and *Ochromonas* requiring thia-mine + biotin.

† 1 *Gymodinium* requiring thiamine + biotin.

requiring B_{12}; producing thiamine and biotin); a minor bloom of flagellates, mainly chrysomonads (predominantly thiamine requirers; some require biotin also); a summer growth of dinoflagellates (predominantly B_{12} requirers); in many localities the summer dinoflagellate growth is scant and an early fall, lesser bloom of diatoms (predominantly B_{12} requirers) may occur.

Obviously, generalizations of this type may be quite misleading since single species of any algal group deviate quite widely from the general trend (compare Table I with Table VIII, pp. 194–95 in Provasoli, 1963). The hypothesis fits also in the well studied succession of species in the Sargasso Sea—a nutrient poor environment, hence the algal growth and vitamin content are quite low. The annual maximum of vitamin B_{12} (0.07 ng/1) is in April and is followed by a small diatom peak and by a drop to 0.01 ng/1 in B_{12} (Menzel and Spaeth, 1962). Most of these diatoms require B_{12} in the laboratory (Guillard and Cassie, 1963), hence should produce thiamine and biotin. The diatom bloom is followed by growth of *Coccolithus huxleyi*, a thiamine-requirer which is endemic the year around, confirming that some thiamine was present (no bioassay for thiamine was done). The small but sustained growth of *C. huxleyi* may account for the slow accumulation of B_{12} from summer to winter.

COMPLEXITIES OF THE B_{12} CYCLE

a) Specificity

Cyanocobalamin (the usual laboratory form of B_{12}; also an anti-pernicious anaemia factor for humans) is one of a family of cobalamins found in nature. The scarce data indicate that from 25 to 70% of marine bacteria produce cobalamins and that bacteria produce far more analogs than cyanocobalamin (10–25% of the total cobalamin); similarly the analogs seem to predominate in seawater (Burkholder and Burkholder, 1956).

The vitamin B_{12}-requiring organisms have 3 characteristic patterns of response. Some utilize cyanocobalamin and other cobalamins with a benzimidazole nucleotide ("mammalian" pattern); the majority of chrysomonads and dinoflagellates have this specificity. The organisms of the "lactobacillus" pattern utilize the cobalamins of the mammalian pattern and the analogs with an adenine-like nucleotide; some dinoflagellates, very few diatoms and chrysomonads have this specificity. The organisms of the "coli" pattern utilize all cobalamins including Factor B which lacks the nucleotide portion of the molecule; so far several diatoms, a few cryptomonads and one blue-green algal species have this pattern of utilization. Organisms having the "coli" pattern have an obvious advantage over the other cobalamin-requirers of narrower specificity. It is perhaps fortuitous that the diatoms, the most productive of three ecologically important algal groups, have more species displaying the "coli" pattern.

Since the ratio analogs: cyanocobalamin seems to vary in seawater and since the algae differ in their specificity, cobalamin specificity may be another relevant nutritional differential. Unfortunately no assessment can be made of the ecological value of this parameter because very few measurements on seawater have been done with two organisms; i.e. one responding only to cyanocobalamin and one responding to all cobalamins (= total cobalamins). The usual bioassay organism for cobalamins is now *Cyclotella nana* (13-1) which gives an intermediate cobalamin value because the growth response to equal concentrations of Factor B and pseudo-B_{12} is, respectively, 20% and 40% of the yield obtained with cyanocobalamin (Guillard, 1968). Similarly, since the excretion of cobalamins by marine algae was followed by Carlucci with the *C. nana* assay, we don't know what type of cobalamin is produced by the algae—a serious gap in knowledge if algae, as postulated, are major producers of vitamins.

b) The B_{12} binding factor

To complicate matters, new findings of Droop (1968) confirm and extend an aspect of B_{12} physiology which might be ecologically relevant. *Euglena gracilis*, a freshwater flagellate requiring B_{12}, liberates in the culture medium a material which binds vitamin B_{12} and makes it unavailable to itself and bacteria alike (Kristensen, 1956). *Ochromonas malhamensis*, a freshwater B_{12}-requiring chrysomonad, behaves similarly (Ford, 1958). Droop (1968) now reports that two marine chrysomonads, *Monochrysis lutheri* and *Isochrysis galbana*, and the marine diatom *Phaeodactylum tricornutum* also produce a B_{12} binding factor. This factor is heat labile, can be separated by gel filtration on Sephadex, and inhibits growth by preventing B_{12} uptake, a competitive inhibition that can be overcome by increasing the concentrations of B_{12} in the medium. The B_{12}-binding factor seems to be non-species-specific; it is released apparently at a constant relative rate by the cells and accumulated in the medium where it competes with the organism for the vitamins.

The results of *Phaeodactylum* are provocative. *Phaeodactylum* does not require B_{12}, therefore its growth is not inhibited by the B_{12}-binding factor, yet it produces the binding factor. Experiments with radioactive B_{12} revealed that *Phaeodactylum*, even though it does not need B_{12}, takes it up at about the same rate as *Monochrysis* or *Isochrysis* which need B_{12}. The binding factor of *Monochrysis*, while not inhibiting growth of *Phaeodactylum* does inhibit B_{12} uptake, revealing the mechanism of action of the inhibitor. The results with *Phaeodactylum*, that a non-requirer takes up B_{12}, conflict with the result obtained with *N. occulata* (Droop, 1968, p. 715) which, like *Phaeodactylum*, does not need vitamins and in syntrophic experiments supplied B_{12} to *Monochrysis lutheri*. [A generalization extending to all algae the finding that a non-requirer takes up B_{12} and that 4 algae produce binding factor would lead to the absurd conclusion that the entire algal biomass consumes B_{12} or makes it unavailable while only a part of the algal biomass products it.]

Profiting from the quantitative data obtained with the chemostat and with the use of ^{57}Co-labelled vitamin B_{12}, Droop (1968) computed the events of the entire growth curve in batch cultures including the production of the binding factor. Logarithmic production of binding factor started at 2/3 along the logarithmic phase of growth when about 9/10 of the B_{12} had already been taken up by *Monochrysis*. Therefore the inhibition effect of the binding factor on the species producing it is minimal—*Monochrysis* had already accumulated enough B_{12} for several divisions. As Droop concluded, a dominant species, by producing the binding factor could effectively maintain its dominance by rendering B_{12} unavailable to other B_{12}-requiring species. The exclusion of B_{12} requirers through the binding factor would favor greatly the thiamine-biotin requiring flagellates by eliminating consumers of N, P, etc. The advent of these flagellates is already favored by production of the two vitamins by the spring bloom of the diatoms. Another role of the inhibitor could be to preserve B_{12} at least temporarily—free B_{12} could be released later from the complex by bacterial action.

BACTERIA–ALGAE RELATIONSHIPS

Even though the bacteria may not be the predominant producers of vitamins in seawater, they certainly are very important producers and consumers of vitamins. Their role will be most active whenever dead organic matter is present as in the decay of large phytoplankton blooms and of the millions of tons of the littoral seaweeds, in polluted areas, in marshes, and bottom muds.

Due to their well-known biochemical versatility contributions of bacteria to relationships between aquatic organisms should be most varied despite the apparent lack of information. It is generally conceded that many bacterized cultures of single algae may grow better, have a "normal" morphology and last longer than the corresponding bacteria free culture, but I am not aware that better growth in bacterized culture was related to specific bacterial products.

It has been reported that blooms of freshwater blue-green algae might be stimulated under conditions of low CO_2 by the CO_2 produced by the abundant bacterial flora colonizing the gelatinous masses and sheets of the blue-greens. Such CO_2 depletion could occur daily at mid-day in superficial blue-green blooms which are often a layer several inches thick of actively photosynthesizing protoplasm. Nutritionists preoccupied with pollution, N, P, trace metals, vitamins, etc. are most prone to forget that CO_2 is the only C source of photoautotrophs.

Seaweeds are also coated by mucilages which harbor "epiphytic" bacteria. This epiphitism may conceal a relationship akin to symbiosis. The mucilages, in slowing down diffusion of the nutrients produced by the algae are a protective niche in which exchange of secretions between bacteria and algae is favored. A specialized microecosystem of species may evolve through the selective action of the antibacterial products of seaweeds as well as from the nature of the algal metabolites, some of which are unusual and can be utilized only by a few enterprising bacteria. Perhaps it is fortuitous that when the juice of ground *Cyclotella nana* (a B_{12}-requiring diatom) is added as a selective enrichment to seawater agar, most of the marine bacteria isolated are B_{12} producers; these bacteria and *C. nana* can be grown syntrophically on media lacking C sources and B_{12} (K. Haynes, personal communication).

Equally interesting is the loss of the normal leafy morphology of *Monostroma oxyspermum* when grown in the absence of bacteria and in mineral media. The normal morphology of *Monostroma* can be restored by the supernatant of 2 marine bacteria isolated from the thallus of *Monostroma* and *Callithamnion* collected from nature. But normal morphology can also be restored by filtrates of several bacteria free curltures of brown and red algae normally found in the same littoral zone—a new example of a binary relationship (bacteria/*Monostroma* or *Monostroma*/other seaweeds) based on factors governing morphological integrity (in preparation; partial accounts in Provasoli and Pintner, 1968 and Provasoli, 1969). Possible effects of algal bloom on growth and composition of the bacterial flora are discussed by Sieburth (1968).

ALGAE AND CRUSTACEA

In establishing bacteria-free cultures of *Tigriopus japonicus*, a marine harpacticoid crustacean, several algae were tried as food organisms. Out of 20 species tried only 6 species of algae satisfied all the nutritional requirements of the predator, allowing growth from new born to fertile adults. Only 1 of these 6 algae (*Monochrysis lutheri*) supported 27 aseptic generations of *Tigriopus* (discontinued accidentally). Severe nutritional deficiencies leading to reduced size of the copepods and infertility occurred after 2 to 9 generations of the predator grown on the other five algae (Provasoli *et al.*, 1959). Since the deficiencies appeared only after several aseptic generations it seemed improbable that gross nutritional imbalance of macronutrients were involved; lack of micronutrients was suspected. Various nutrients were added singly and in combination to the media in which the algae and *Tigriopus* were growing. A vitamin mixture restored fertility for 4 generations or more when the food organism was either *Isochrysis* or *Chroomonas* (Shiraishi and Provasoli, 1959).

A similar case developed when D'Agostino tried to grow the amphigonic and parthenogenetic races of *Artemia salina* at different salinities with *Dunaliella viridis* and *D. salina* as food organisms. The two races of *Artemia* grew well at all salinities above seawater concentration, at 3% or lower they failed to develop to adults or the adults were infertile. Addition of liver and yeast extract (0.5 mg% each) restored normal development and fertility (D'Agostino and Provasoli, 1968).

More recently, *Daphnia magna* fed on *Scenedesmus obliquus* and *Chlamydomonas reinhardi* in mineral media supporting the growth of the algae, lost fertility after 10 aseptic generations. Fertility was restored by adding to the mineral medium 7.6 mg% of peptone and 0.5 mg% of yeast extract. In all these cases the food algae used (except *Isochrysis* which needs B_{12} and thiamine) require for growth neither vitamins nor organic compounds; they are photoautotrophic. Since the organic enrichments were added to the mineral media used for growing together the algal food and the crustacea, it was not clear whether the enrichment acted on the crustacean or *via* the algae. Indirect evidence indicated that the enrichment acted via the algae making them more nutritious (see discussion in D'Agostino and Provasoli, 1968).

This was finally proved when *Scenedesmus* and *Chlamydomonas*, the algal food for *Daphnia*, were grown separately on agarized inorganic medium and on the organic enriched agarized medium. The algae were then carefully scraped off the agar and fed as necessary to *Daphnia* in the liquid mineral medium—all the operations were aseptic. *Daphnia* fed on algae grown on mineral media lost fertility and died at the 12th generation; *Daphnia* fed on algae grown on the organic enrichment maintained fertility until the 20th generation (discontinued). Experiments in progress indicate the possibility that an enrichment constituted of B_{12}, thiamine and pantothenic acid is as effective for fertility maintenance as the whole organic enrichment.

Since vitamins also restored fertility to *Tigriopus*, how can we relate them to the increased nutritional value of the algae so treated for the crustacea? It is well known that algae accumulate all sorts of substances and the vitamin effect could be simply due to this. However there is an indication that the metabolism of some algae is affected by exposure to B_{12} even if not needed for growth, resulting in changes in cell composition, such as increased proteins, fats and other vitamins (see discussion in Provasoli, Conklin and D'Agostino, 1970).

Therefore the content in organic solutes and vitamins of waters may affect in an important way the nutritional value of the algae for the predator—an unsuspected relationship mediated by the extracellular products.

POSTSCRIPT

In the oral presentation two relationships were presented and documented: a) the action of bacterial and seaweed supernatants in restoring the normal morphology of *Monostroma oxyspermum*, b) the effect of vitamins and organic solutes on the nutritional value of the algae for predators. Since these phenomena were presented at two previous Symposia (Provasoli, 1969 and 1970) they are included here in an abbreviated form.

The present exposition is based on very few but provocative data which suggest the necessity of more investigations on vitamin cycles. The hypotheses expressed may be attractive to some and outrageous to others; in either case it is hoped that they may stimulate more work. Studies on the B_{12} binding factor and on morphogenetic factors in other seaweeds are our present interest.

5*

A concern for interactions seems important because it may lead to the identification of new parameters, hence more detailed models. Phytoplankton productivity based on ^{14}C uptake has served well in the preliminary phase of describing the flow of energy in the system, but without simultaneous data on the species composition, it does not allow correlation of environmental parameters with a defined biocenosis and its changes. A surge in ^{14}C uptake may not result in higher productivity of herbivores and carnivores if the increased algal population is unacceptable in size or taste, is toxic or lacks some necessary nutrients.

During the last 20 years the focus has been on the lower (phytoplankton) and upper (fishes) thirds of the food pyramid; attention should now be given to the middle third to define the intricacies of the food web. Meanwhile aquaculture is rapidly growing in Japan and elsewhere and needs much basic input to avoid repeating in the sea the sad experiences of the last century in the unwise use of the land and lakes.

Acknowledgement

I am most grateful to Dr. Marta Vannucci, the organizing committee, and the Institutions supporting the International Symposium for a splendid organization and their support for travel. The reas2rch was supported by contract NR 104-202 with the Office of Naval Research and research grant GB-12078 of the National Science Foundation.

References

BARBER, R. T. and RYTHER, J. H. (1969). Organic chelators: Factors affecting primary production in the Cromwell Current upwelling. *J. Exp. Mar. Biol. Ecol.* 3, 191–99.

BEDNAR, T. W. and HOLM–HANSEN, O. (1964). Biotin liberation by the lichen alga *Coccomyxa* sp. and by *Chlorella pyrenoidosa. Plant Cell Physiol.* Tokyo, 5, 297–303.

BURKHOLDER, P. R. (1963). Some nutritional relationships among microbes of sea sediments and water. In *Symposium on marine microbiology* pp. 133–150 (ed. Oppenheimer, C. H.) C. C. Thomas, Springfield, Ill.

BURKHOLDER, P. R. and BURKHOLDER, L. M. (1956). Vitamin B_{12} sin suspended solids and marsh muds collected along the coast of Georgia, *Limnol. and Oceanogr.* 1, 202–208.

BURKHOLDER, P. R. and BURKHOLDER, L. M. (1958). Antimicrobial activity of horny corals. *Science* 127, 1174–1175.

BURKHOLDER, P. R. and BURKHOLDER, L. M. (1958). Studies on B vitamins in relation to the productivity of the Bahia Fosforescente, Puerto Rico. *Bull. Mar. Sci. Gulf and Carribean* 8, 201–223.

CARLUCCI, A. F. and SILBERNAGEL, S. B. (1966, 1967). Bioassay of seawater I, II, IV. *Can. J. Microbiol.* 12, 175–183, 1079–1089; 13, 979–986.

CARLUCCI, A. F. (1969). Vitamin production by algae. *Abst. 32nd meeting of the Amer. Soc. of Limnol. and Oceanogr.*

CRAIGIE, J. S. and McLACHLAN, J. (1964). Excretion of colored ultraviolet-absorbing substances by marine algae. *Can. J. Bot.* **42**, 23–33.

D'AGOSTINO, A. S. and PROVASOLI, L. (1968). Effects of salinity and nutrients on mono- and diaxenic cultures of two strains of *Artemia salina. Biol. Bull.*, Woods Hole. **134**, 1–14.

DROOP, M. R. (1968). Vitamin B_{12} and marine ecology. IV. The kinetics of uptake, growth and inhibitions in *Monochrysis lutheri. J. Mar. Biol. Assoc. U. K.* **48**, 689–733.

ERICSON, L. E. and LEWIS, L. (1953). The occurrence of vitamin B_{12} factors in marine algae. *Arkiv Kemi.* **6**, 427–442.

FOGG, G. E. (1962). Extracellular products. In *Physiology and biochemistry of algae.* pp. 475–492. (ed. Lewin, R. A.) Academic Press, N. Y.

FOGG, G. E. (1966). The extracellular products of algae. *Oceanogr. Mar. Biol. Ann. Rev.* **4**, 195–212.

FOGG, G. E. (1969). The Leeuwenhoek lecture 1968. The physiology of an algal nuisance. *Proc. Roy. Soc. B.* **173**, 175–189.

FOGG, G. E. and WESTLAKE, D. F. (1955). The importance of extracellular products of algae in freshwater. *Proc. Inter. Assoc. Theor. Appl. Limnol.* **12**, 219–232.

FORD, J. E. (1958). B_{12}-vitamin and growth of the flagellate *Ochromonas malhamensis. J. gen. Microbiol.* **19**, 161–172.

FORD, J. E. and GOULDEN, J. D. S. (1959). The influence of vitamin B_{12} on the growth rate and cell composition of the flagellate *Ochromonas malhamensis. J. gen. Microbiol.* **20**, 267–276.

GOLD, K. (1964). A microbiological assay for vitamin B_{12} in seawater using radiocarbon. *Limnol. Oceanogr.* **9**, 343–347.

GUILLARD, R. R. L. (1968). B_{12} specificity of marine centric diatoms. *J. Phycol.* **4**, 59–64.

GUILLARD, R. R. L. and CASSIE, V. (1963). Minimum cyanocobalamin requirements of some marine centric diatoms. *Limnol. Oceanogr.* **8**, 161–165.

HELLEBUST, J. A. (1965). Excretion of some organic compounds by marine phytoplankton. *Limnol. and Oceanogr.* **10**, 192–206.

JOHNSTON, R. (1964). Seawater, the natural medium for phytoplankton. II. *J. Mar. Biol. Assoc. U. K.* **44**, 104–116.

JORGENSEN, E. G. (1962). Antibiotic substances from cells and culture solutions of unicellular algae with special reference to some chlorophyll derivatives. *Physiol. Plant* **15**, 530–545.

KRISTENSEN, H. P. O. (1956). A vitamin B_{12}-binding factor formed in cultures of *Euglena gracilis.* var. *bacillaris. Acta. Physiol. Scand.* **37**, 8–13.

LÉFÈVRE, M., JAKOK, H. and NISBET, N. (1952). Auto- et héteroantogonisme chez les algues d'eau douce. *Ann. Stat. Centr. Hydrobiol. Appl.* **4**, 197 pp.

LEWIN, R. A. (1956). Extracellular polysaccharides of green algae. *Can. J. Microbiol.* **2**, 665–672.

LUCAS, C. E. (1947). The ecological effects of external metabolites. *Biol. Rev.* **22**, 270–295.

MENZEL, D. W., HULBURT, E. M. and RYTHER, J. H. (1963). The effect of enriching Sargasso sea water on the production and species composition of phytoplankton. *Deep-Sea Res.* **10**, 209–219.

MENZEL, D. W. and RYTHER, J. H. (1969). Distribution and cycling of organic matter in the oceans. *Proc. Symp. Organic matter in natural waters.* Univ. of Alaska (in press).

NAKAMURA, K. and GOWANS, C. S. (1964). Nicotinic acid excreting mutants in *Chlamydomonas. Nature*, London. **202**, 826–827.

NIGRELLI, R. F. (1958). Dutchman's "baccy juice" or growth promoting and growth-inhibiting substances of marine origin. *Trans. N. Y. Acad. Sci.* Ser. II, **20**, 248–262.

PRATT, R., MAUTNER, H., GARDNER, G. M., SHA-YI-HSIEN and DUFRENAY, J. (1951). Antibiotic activity of seaweed extract. *J. Amer. Pharm. Assoc.* (Sci. ed.) **40**, 575–579.

PROVASOLI, L. (1963). Organic regulators in phytoplankton fertility. *The Sea* Vol. *II*, pp. 165–219 (ed. Hill, M. N.) Interscience N. Y.

PROVASOLI, L. (1969). Algal nutrition and Eutrophication. In *Eutrophication: causes, consequences, correctives. Nat. Acad. Sci. Washington, D. C.*

PROVASOLI, L., CONKLIN, D. E. and D'AGOSTINO, A. S. (1970). Factors inducing fertility in aseptic Crustacea. *Helgolander wiss. Meeresunters.* (in press).

PROVASOLI, L. and PINTNER, I. J. (1953). Ecological implications of *in vitro* nutritional requirements of algal flagellates. *Ann. N. Y. Acad. Sci.* **56**, 839–851.

PROVASOLI, L. and PINTNER, I. J. (1964). Symbiotic relationships between microorganisms and seaweeds. *Amer. J. Bot.* **51**, 681.

PROVASOLI, L., SHIRAISHI, K. and LANCE, J. R. (1959). Nutritional idiosyncrasies of *Artemia* and *Tigriopus* in monaxenic culture. *Ann. N. Y. Acad. Sci.* **77**, 250–261.

SHIRAISHI, K. and PROVASOLI, L. (1959). Growth factors as supplement to inadequate algal foods for *Tigriopus japonicus. Tohoku J. Agr. Res.* **10**, 89–96.

ROBBINS, W. J., HERVEY, A. and STEBBINS, M. (1951). Further observations on *Euglena* and B_{12}. *Bull. Torrey Bot. Club* **78**, 363–75.

RYTHER, J. H. and GUILLARD, R. R. L. (1962). Studies of marine planktonic diatoms: II. Use of *Cyclotella nana* for assays of vitamin B_{12} in seawater. *Can. J. Microbiol.* **8**, 437–445.

SIEBURTH, J. McN. (1959). Antibacterial activity of Antarctic marine phytoplankton. *Limnol. and Oceanogr.* **4**, 419–424.

SIEBURTH, J. McN. (1968). The influence of algal antibiosis on the ecology of marine microorganisms. In *Adv. Microbiol. Sea*, pp. 63–94 (eds. Droop, M. R. and Ferguson Wood, E. I.) Academic Press, N. Y.

SIEGEL, A. (1967). A new approach to the concentration of trace organics in seawater. In *Pollution and Marine Ecology*, pp. 235–256 (eds. Olson, T. A. and Burgess, F. I.) Interscience N. Y.

SMAYDA, T. J. (1970). Growth potential of water masses using diatom cultures: Phosphorescent Bay (Puerto Rico) and Caribbean Waters. *Helgolander wiss. Meeresunters.* (in press).

STRICKLAND, J. D. H. (1965). Productions of organic matter in primary stages of the marine food chain. In *Chemical Oceanogr.* Vol. *I*, pp. 477–610 (eds. Riley, J. P. and Skirrow, G.) Academic Press, N. Y.

THOMAS, W. H. (1969). Phytoplankton nutrient enrichment experiments off Baja California and in the Eastern Equatorial Pacific Ocean. *J. Fish. Res. Board Canada* **26**, 1133–1145.

ZEHNDER, A. (1949). Über den Einfluß von Wuchsstoffen auf die Flechtenbildner. *Ber. Schweiz. bot. Ges.* **59**, 201–267.

Alternative sources of food in the sea

J. E. G. RAYMONT

Department of Oceanography, The University of Southampton,
England

Resumo

Alternativas à cadeia alimentar clássica: produção primária, produção secundária, carnívoros primários. O papel dos quimioautróficos. O uso de bolotas fecais, e produtos de degradação de clorofila. Valor nutritivo das fezes. Matéria orgânica particulada em suspensão. Quantidade e significância nas camadas superiores e em águas profundas. Possível uso como alimento, e como substrato para bactérias. Uso dos detritos diretamente como alimentação por animais bentônicos e por plâncton. Existência de fitoplâncton em grandes profundidades e sua heterotrofia.

Matéria orgânica dissolvida; sua predominância como parte da matéria orgânica total. Frações "estáveis" e "instáveis". O problema da nutrição das bactérias. A magnitude e rapidez do fluxo de matéria orgânica dissolvida. O uso direto, de matéria orgânica dissolvida por Protozoários.

O uso indireto de matéria orgânica dissolvida. Sua conversão em agregados orgânicos, e alimento em forma de matéria particulada. Diferenças entre áreas costeiras e interiores e profundidades oceânicas.

Comparação de produção primária e heterotrofia. A importância de requisitos de dieta suplementares, no ambiente marinho: Vitaminas e substâncias de crescimento. Secreção de matéria orgânica dissolvida, utilizável como fontes de energia. Possível resíntese de materiais secretados.

Various trophic pathways have been proposed in the ocean, in addition to the usual route of consumption of phytoplankton by herbivorous plankton, and consumption of herbivores by carnivorous zooplankton. Some of the suggestions have arisen from the problem of relating the amount of primary production to the crop of consumers. Although at times phytoplankton production appears to meet grazing requirements, over certain periods the herbivores would appear at best to be existing at a low food level (e.g. Adams and Steele, 1966).

Chemo-autotrophic bacteria have been considered as supplying an additional food source. Though such organisms occur in the marine environment, they are not common and appear to be abundant only in regions where H_2S concentration is high; even then they are restricted to certain zones in the oceans. For example, Sorokin (1964) cites these autotrophs as occurring only in the boundary between the oxygenated and deoxygenated layers in the Black Sea. In any event such chemo-autrotrophs utilize compound (H_2S) formed under reducing conditions typical of high organic accumulation, and thus to that extent are dependent on the primary production of some past time period. Chemo-autotrophy may be considered to contribute little to the productivity of oceans today, though possibly of greater importance in the primaeval aquatic environment.

The problem of alternative food sources is most acute in the deep sea (cf. Nishizawa, 1966). Until comparatively recently the deeper living zooplankton animals were regarded either as filter feeders, dependent primarily on a thin rain of particles descending from the euphotic zone, or as carnivores preying upon these filter feeders. Considerable doubts have been expressed about the nutritive value of such particles (Menzel and Goering, 1966; Riley, Wangersky and Van Hemert, 1965) but several workers (e.g. Johannes and Satomi 1966) have proposed that faecal pellets, though included amongst such particles, were of special significance as a food for zooplankton. Vinogradov (1962) believes that zooplankton during their vertical migrations fed in the upper layers and then descended, bringing material including faecal pellets with them. In this way, ladders of feeding chains proceeding down into the depths of the ocean might be set up.

There have been few quantitative assessments of the nutritive value of such pellets. But a possible lead may stem from Newell's (1965) observations on a benthic gastropod, *Hydrobia*. Newell found that this animal would ingest and partially assimilate its own faeces. A more important observation, however, was that the faeces first produced were relatively high in carbon but very low in nitrogen, and were virtually unacceptable as food. After three days there was a great increase in nitrogen (approximately × 80) although the carbon was slightly reduced. Newell presumed that the freshly voided faeces were used as a surface for microheterotrophs, largely voided from the gut of the feeders. These heterotrophs used the carbon in the faeces to some extent as a source of energy, but more importantly, they absorbed nitrogen from the dissolved organic matter in the sea water and increased their own body protein. Thus the bacteria increased the food, especially protein, available to *Hydrobia*.

A rather similar study by Johannes and Satomi (1966) showed that *Palaemonetes pugio* also voided faeces which were reaccepted. The faeces were densely packed with bacteria of intestinal origin; some 20% of the material was organic carbon, with a predominance of protein. There was a high assimilation efficiency of the protein carbon available. The micro-organisms in the faecal material could incorporate some dissolved organic matter into particulate body substance though apparently only at relatively high concentrations. If Newell's (1965) and Johannes and Satomi's (1966) observations on the faeces of benthic animals can be translated to the plank-tonic marine environment, there is the possibility that marine micro-organisms may make increased use of detrital products. If they also can utilize dissolved organic matter and produce protein material, additional food would be available for use by filtering animals. In Johannes' words, this could reduce entropy in the marine foodweb.

In the open oceans faecal pellets often tend to sink relatively faster than some of the more loosely aggregated general organic material derived from the euphotic zone. Some Crustacea, for example, produce discrete feacal pellets, some wrapped in a peritrophic membrane; their rate of descent to the lower layers is thus more rapid, so that the loss of the "useful" material is reduced.

Thus workers (e.g. Vinogradov, 1962; Banse, 1964) have suggested that a variety of deeper living zooplankton, including decapods, euphausids, copepods, mysids, ostracods and perhaps some protozoans feed partly on the faecal remains of other zooplankton animals. There are likely to be variations in the quality of nutriment in the faeces of different plankton species. In some forms the plant material would appear to have suffered relatively little decomposition. Thus Nemoto (1968) has recently investigated the chlorophyll pigments in the stomachs of several euphausids. A near surface feeder such as *Euphausia similis* shows stomach contents having the highest percentage of undegraded chlorophyll pigments of the species investigated, and also produces an abundance of faecal pellets. By contrast, the stomach contents of deeper living forms, such as *Benthophausia*, contain chlorophyll pigments which are mainly degraded. Nemoto suggests that these deeper living species may feed on "phyto-detritus" and also on the voided faecal pellets during the vertical migration of other species.

Faecal material and all plankton residues tend to break down into smaller particles and occur as organic particulate suspended matter (detritus). Even in the euphotic zone the dead material forms a significant fraction (c.f. Krey, 1961; Hagmeier, 1964). Although the uppermost layers contain distinctly larger amounts of detritus than deeper waters, and though there are considerable regional and temporal variations in the surface strata, all workers seem now agreed that there is a fairly rapid reduction in the first few hundred metres. Below a depth of some 300–500 m the amount of particulate organic matter is relatively unchanged. For open oceans detritus in the uppermost layers is usually of the order of 0.1 mg C/litre (Menzel, 1967) though a few higher values have been reported (e.g. Skopintsev, 1965). The fall to depths of 300–500 metres is relatively very sharp. Menzel and Ryther (1968) give values for the S. W. Atlantic which suggest that below 400 metres particulate organic carbon is only of the order of 5 μg C/l. Szekielda's (1968) data for deeper layers of the North Atlantic range from 60–120 μg C/l; Riley, Van Hemert and Wangersky (1965) found for the Sargasso Sea concentrations usually between 10 and 20 μg C/l layers below 500 m. Most authors agree that the variations in deep water are small and appear to be random. Strickland (1965), however, while recognising these low values for detritus in deeper layers, points out that this "is a significant food supply for zooplankton at all depths remembering that under a square metre of surface in the open oceans, as compared with some 2–3 grams standing crop of phytoplankton there will be some 100 grams of detritus. Even though this mostly consists of persistent animal remains such as chitin, the detritus must enter presumably into the food chain".

The amount of detritus in the deeper oceans is not only very small; the sedimentation rates of these fine particles is also exceedingly slow. Rates have been suggested ranging from 1 to 7 metres per day, and at such speeds, the sedimentation time to the bottom of the average deep ocean might take from one to ten years. Presumably extensive decomposition of the material must have proceeded over this period so that the nutritive value of such detritus would be negligible. Portions of cellulose, lignin, chitin and other materials of a relatively resistant nature are common in such deep sea detrital material. The detritus is thus small in quantity and low in nutrient material. Nevertheless, Wheeler (1967) has recently shown that "carcasses" of zooplankton are identifiable at considerable depths, and though largely chitin, the content of such dead matter equals or perhaps exceeds the crop of zooplankton at such depths.

The small amount of nutritious food may be correlated with the tremendous reduction in populations experienced at great depths. Zenkevitch and Birstein (1956) have suggested that the biomass of both deep sea plankton and of benthos is very reduced. Vinogradov (1962), commenting on the extraordinary rapid reduction of biomass of zooplankton in really deep water, suggested that at high latitudes the biomass below 6000 metres is only 1000 th that of the uppermost water layers. Banse (1964) also gives data showing reduction of 2–3 orders of magnitude in deep water. Skopintsev (1965) quotes data from Yashnov; the vast proportion of the zooplankton in the open oceans is in the upper 1000 m and especially the 0–500 m level. Recent investigations on a decrease in biomass of benthos of 3 orders of magnitude from surface to 5000 m have already been described in this Symposium (Reference: Rowe, G. T. "Benthic biomass and surface productivity"). This reduction in biomass of deeper layers applies apparently to all elements of the marine ecosystem—bacteria, benthos as well as zooplankton (cf. Skopintsev, 1965).

Some deposit feeders amongst the deep sea benthos (e.g. Holothurians) probably a certain amount of nutriment directly from the detritus despite its low nutritive value. An important use, however, of this detritus may be to act as a surface for bacteria, the micro-organisms gaining nutriment by absorbing dissolved organic matter. Some bottom invertebrates from shallow inshore areas undoubtedly utilize bacteria as food. Many investigators have shown that a considerable proportion of the food of such species comes from this source. Bacteria may presumably be equally utilized by deep sea benthos but it would be impossible to separate detritus from bacteria as a foodstuff.

One must particularly consider here the meiobenthic fauna especially small copepods and some nematodes. Recent investigations by MacIntyre, Munro and Steele (1968) employed sand columns in the laboratory which developed a rich meiofauna of harpacticoid copepods and nematodes. This meiofauna very successfully utilised the bacterial food produced in the experimental conditions. The bacteria undoubtedly gained most of their substrate from dissolved organic matter adsorbed on the sand surfaces.

It is more difficult to resolve how far the zooplankton can utilise bacteria. Strickland (1965) states that Zhukova, and Voroschilova and Dunova have found success in use of bacteria by zooplankton. On the other hand, the early experiments of Clarke and his colleagues on copepods (e.g. Fuller and Clarke, 1936) and the numerous experiments on feeding bivalve larvae by Loosanoff and his school (e.g. Davis, 1953; Loosanoff and Davis, 1963) and also by Walne (1958), all suggest that bacteria are not suitable as food for zooplankton. Possibly it is not fair to compare these coastal shallow forms with deep sea zooplankton. A recent interesting contribution by Takahashi and Ichimura (1968), dealing with fresh waters, may be significant in this connection. These workers showed that sulphur bacteria were definitely consumed by copepods just above the anoxic zone. Bacteria were labelled and the copepods were found to possess the tracer. Perhaps deep sea marine filterers may act similarly.

We are ignorant of even the outlines of trophic pathways from detritus in the oceans. A possible food chain from detritus through bacteria might involve Protozoa. Murray (1963) suggested that Foraminifera might feed on detritus and on bacteria. However experiments suggest that living phytoplankton is preferred. Other recent work with labelled cultures strongly suggest Foraminifera prefer phytoplankton; bacteria though ingested did not contribute apparently to the nutrition. Radiolaria may be abundant, some chiefly in deep water, and are known to be food for a number of deep sea zooplankton forms. How far Radiolaria, however, may feed on bacteria is unknown. Ciliates are sometimes found in some abundance in sea water, and their rate of turnover is high when compared with most Metazoa (cf. Steemann Nielsen, 1962). They are also frequently encountered in laboratory cultures. However, their numbers and rates of reproduction as compared with bacteria are low. By comparison with freshwater species, marine infusoria might be expected to feed at least to some degree on bacteria, but direct evidence is sparse. Zaika and Averina (1968) have found in the admittedly shallow waters at the entrance to Sevastopol Bay reasonable densities of infusorians. "Small" ciliates (20–55 μ) showed a mean density over four summer months of 1600/l and "large" species (> 55 μ) some 350/l. Several genera were recognised but no evidence is available on their feeding activities. Gold (0000) and other workers have also recently cultured Tintinnids. These ciliates, however, appear to feed on phytoplankton and preferentially on dinoflagellates. On the other hand, Lighthart (1969) working in the Puget Sound area, was able to isolate a few species of bacteriovorous Protozoa, mostly flagellates, from offshore waters in the water column; higher densities of both Protozoa and bacteria were found inshore, especially in the bottom sediments.

One of the main difficulties in envisaging a food chain from detritus through bacteria has been the emphasis that the density of bacterial populations in the deep sea is so low that they can have very little nutritional value. Investigations of Holm–Hansen and Booth (1966), using ATP as an indicator of biomass of living biological material, (cf. also Holm–Hansen—this Symposium) have shown that the mean microbial population for a column from below the euphotic zone (ca. 200 to 1000 metres depth) was about 1×10^6 or more per litre—many times higher than densities cited by most microbiologists for plating counts. The deficiencies of plating counts in estimating true bacterial populations are now, however, widely recognised. The more recent investigations suggest that the deep sea, while not having in anyway a rich bacterial population, may have a density which could help to support bathypelagic zooplankton.

Skopintsev (1965) suggests that the excessively small quantities of particulate suspended matter at great depths are still far larger than those deduced from maximum sedimentation rates. He has suggested that this is a reflection of the importance of detritus in the feeding of deep sea organisms. But lateral transport of particles can greatly affect sedimentation rates and it is doubtful how far this can be used as a proof of the direct use of detritus. Banse (1964) strongly advocates the lateral transport of "rich" water such at A.I.W. bringing relatively large quantities of detritus to other regions. This view is supported by Szekielda's (1968) observations: surface productive areas at high latitudes are frequently marked by the sinking of waters, especially in the Antarctic; the bottom water is then rich in particulate carbon. Even as far north as in the Tasman Sea bottom water showed the relatively high concentration of $>200\ \mu g\ C/l$.

The work of Bernard (1948, 1953, 1959, 1963, 1967) and of Wood (1959, 1963, etc.) in conjunction with several other workers has drawn attention to the existence of relatively deep living phytoplankton. It is clear from Bernard's recent review that in the Mediterranean coccolithophores are by far the commonest organisms, and are often at least as rich in the deeper layers as they are in the euphotic zone. *Nostoc* and a few other Myxophyceae are also present. Flagellates have been noted by Fournier (1966) in the Atlantic, and the presence of the "olive green" cells has been reported by several workers, including recently by Riley *et al.* (1965) for the Sargasso, where coccolithophores and possibly diatoms also occurred. Recently Strickland and colleagues (1968) have reported green cells in deep water down to 3000 m. Dinoflagellates have also been identified in deep layers.

The total living organic matter represented by these cells is not negligible. Wood has shown both by fluorescence methods and also by culturing that some of the deep living diatoms are viable. There is little doubt then that this deep living plankton is alive, containing chlorophyll. Although phagotrophy has been demonstrated in some Chrysophyceae and dinoflagellates, the likeliest suggestion is that such phytoplankton cells at great depths live heterotrophically on the dissolved organic matter. Such workers as Pintner and Provasoli (1963), Lewin (1963), Provasoli and McLaughlin (1963), Guillard (1963) and Hellebust and Guillard (1967) have demonstrated that a variety of algal groups have heterotrophic abilities.

It is impossible to say how far such a deep living phytoplankton population is perpetuating itself, though Bernard gives a division rate for *Cyclococcolithus* (the commonest coccolithophore at great depths in the Mediterranean) of 1 division in 10 days, which is impressive. How far such a population, even if actively growing heterotrophically, and thus competing with bacteria, can withstand the onslaughts of a grazing population is difficult to tell, but at least such a deep living plankton must be included as a possible minor accessory food source for the very limited bathypelagic population which exists at great depths. There is also a rapid fall of coccoliths to the depths ($\sim$ 1–6 days), according to Bernard, since *Cyclococcolithus* occurs in palmelloid plates, containing $10–180 \times 10^3$ cells, some 2–4 cells thick.

The evidence of zooplankton feeding on phytoplankton in the aphotic zone is admittedly very slight; a deep living ostracod, not found at shallow depths, has been reported with green phytoplankton in the gut. Similarly some of the deeper living euphausids have been noted as having green phytoplankton. It is possible, however, that such material could have arisen from faecal pellet feeding.

Another source of potential nutriment lies in the dissolved organic matter in sea water. Though not very concentrated, the amount is usually an order of magnitude greater than that of the particulate matter. The term "dissolved organic matter" is one of convenience; in practice it is the material that passes through filters in the pore size 0.5–1 μ range. It thus contains some organic material in true solution and some extremely fine particulate material.

Duursma (1963) has suggested from data from the North Sea, Norwegian Sea and North Atlantic an average concentration of dissolved organic matter of 0.8 mg carbon/litre. Mean values which are rather similar have been published for the Mediterranean and for Pacific (Australasian) waters. Menzel and Ryther (1968) present data which are of the same order of magnitude, surface dissolved organic C approaching 1 mg/litre. Skopintsev's (1965) data are somewhat higher; they apparently include the small amount of particulate matter, but differences in analytical procedure may account for much of the variation.

There seems fairly general agreement that though the amount of dissolved organic matter near the surface may be somewhat greater than the mean value, below about 400 metres the variations with depth are relatively insignificant. Menzel and Ryther (1968), for example, suggest that from 400 m to the bottom at ca. 5000 m the organic carbon does not vary very much from 0.5 mg carbon/litre. Obviously, however, the total amount of dissolved organic matter beneath a square metre of surface very greatly exceeds the annual production. This has led numerous workers to emphasize the very conservative nature of dissolved organic matter and that the material must be of considerable age. It is impossible to say with certainty at the present time whether the amount of dissolved organic matter is slowly increasing, but certainly it appears to be extremely resistant to oxidation. It has been spoken of as a "marine humus". Williams, Oeschger and Kinney (1969), while commenting on its great age, have pointed out that at an average of only 0.5 mg C/litre concentration, the reservoir for the world oceans is of the order of $\times 10^{17}$ grams of carbon. The input from the amount of average photosynthesis under a square metre/year fixed by phytoplankton may also be computed; it appears that only about 0.5% of this enters the deep seas as dissolved organic matter. The great resistance of this organic matter to bacterial action and oxidation is therefore apparent. Menzel and Ryther also comment on the relatively unchanging material which composes the dissolved organic matter. Szekielda (1968) emphasises that a major part of the dissolved organic matter in the deep oceans is very resistant to decomposition, and believes that it is derived from surface material which has sunk at high latitudes, or alternatively, from Mediterranean-type seas where there is outflow of rich deep water. The vertical descent of particles from the water in the particular area under observation is only a relatively minor addition to the dissolved organic matter of that area.

Most workers (Strickland, Duursma, Menzel, Skopintsev, Khailov, etc.) are agreed, however, that while the bulk of dissolved organic matter is a relatively stable humus, a small quantity is in a much more labile form.

Khailov (1965) refers to some relatively large molecules composed of poly-saccharides and proteins among the unstable fraction, but the majority of the small amount of labile material is probably represented by smaller molecules. Khailov and Finenko (1968) have recently demonstrated that macromolecules of dissolved organic matter are actively adsorbed on the natural detritus in sea water. Micro-organisms on the surface of the detritus assimilate a portion of the adsorbed macromolecules but part is broken down into dissolved organic matter of smaller molecular weight and is released again to accumulate in the sea water.

There is indeed a most impressive list of specific organic substances found in sea water. Duursma (1965) quotes a variety of carbohydrates, including pentoses, hexoses and traces of rhamnosides, protein derivatives including polypeptides and some twenty amino acids, and a variety of lipids, amongst which are saturated and unsaturated fatty acids of chain length at least from C_{12} to C_{22}, as well as smaller molecule organic acids. This list does not include traces of "biologically active" substances such as vitamins and similar activators, inhibitors, chelators of metals, etc.

Although occasionally values have been quoted particularly for hexoses in excess of 10 µg/litre, in general the quantities of carbohydrates and amino acids range from <1 to about 10 µg/litre. Almost nothing is known of their origin; part might arise from very slow decomposition of the "stable" fraction. A more likely source is the secretions of organisms, but then mainly in or near the euphotic layer. Their importance in trophic pathways, however, is that, though in extremely low concentration, such labile materials will provide food for heterotrophic organisms.

An interesting feature is that there appears to be very little change in the concentration of these labile substances with depth. The standing level of concentration near the surface must, of course, represent the difference between their production in the euphotic zone and their removal. How the balance is maintained and a more or less constant concentration kept throughout the water column is more difficult to interpret. The movement of water masses, particularly those from productive high latitudes, and downwelling processes must play a significant role. But as Khailov (1965) emphasises, it is the rate of interchange which is all important, so that although microbial organisms throughout the water column utilise the la-bile fraction, a more or less constant though low concentration is maintained.

Jannasch's (1967) observations suggest how bacteria may live at such very low dilutions. There appears to be a marked contrast to "batch" bacterial populations, grown in relatively small volumes in the laboratory, and populations in the natural environment. In the laboratory very much higher concentrations of nutrients are required, and a very few species of micro-organisms dominate the cultures. In the sea, at much lower substrate levels, a number of bacterial species seem able to grow in mixed populations. The system appears more to approach a dynamic balance.

There is some evidence that any artificial increase in the amount of easily oxidisable labile organic material in the sea is very rapidly reduced by heterotrophs so that these substances are kept at a relatively low and constant level. The critical factor would appear to be the rate of turnover time rather than the concentration. In this connection, I am greatly indebted to Dr. P. J. le B. Williams of my Department for permission to refer to some of his unpublished data.

At station E1, some 20 miles offshore, in the English Channel, Dr. Williams finds that glucose and aminoacids show pronounced seasonal changes in turnover time. This is in contrast to their overall concentrations which appear to vary relatively little. During the winter in the English Channel the rate of oxidation is low; less than 1 % is oxidized daily. On the other hand, during the summer some 30–50% of glucose is oxidized per day, and a lesser amount, usually 5–10% of amino acids. A few more or less isolated observations in more offshore areas suggest rather similar breakdown rates; in truly oceanic waters the rates are probably of the order of a tenth of those quoted. It would be of considerable interest to study rates in tropical waters.

Some further observations by Dr. Williams in the same region in the upper waters of the English Channel indicate that the total amount of glucose oxidized by heterotrophic organisms during the course of the year amounts to about 6 g/m². Total amino acids are oxidised to the extent of about 30 g/m²/year. Of the amount taken up by heterotrophs, one-third appears to be utilised as substrate for oxidation but as much as two-thirds is incorporated into the cell. Thus some 70 g of bacterial cellular material (say 30 g C) is produced annually per m² of surface from glucose and amino acids. This is not an insignificant fraction of the primary production—some 200–300 g C/m².

The possible trophic routes to Metazoa, once dissolved organic matter has been converted into microbial living material, have already been outlined (cf. page 9). Seki (1966) has described one rather similar food chain in a laboratory cultivation of the brine shrimp.

There remains the possibility that Protozoa might absorb dissolved organic matter directly, rather than feeding on bacteria.

Although some protozoan species absorb nutritive materials in solution in laboratory culture, bacteria would seem at a marked advantage over Protozoa, especially with their adaptability of enzymes, for feeding on particulate but especially on dissolved organic matter. It would be interesting to have a direct comparison of the efficiency of a protozoan and a bacterium for absorbing dissolved organic material. If any parallel may be drawn by a comparison of a fresh water bacterium and alga, the work of Wright and Hobbie (1965) may be pertinent. In their experiments, using glucose and acetate as substrate, they showed that at very low nutrient levels bacteria were at a considerable advantage and used a transport system which was not available to algae. At higher concentrations a diffusion mechanism operated which allowed the algae to absorb the substrate.

Thus in the oceans, microbial heterotrophs are probably by far the most effective organisms in converting dissolved organic matter to living particulate substance, though the great bulk of the dissolved organic matter is relatively stable and is virtually unused in the food chain.

Stephens (1967) has shown for a variety of soft-bodied animals, drawn from several phyla, that direct uptake of dissolved organic matter is possible by metazoan animals. While this might supply traces of certain essential nutrients, there is little likelihood that this mechanism can make any substantial contribution to the normal food supply of Metazoa in the marine environment.

The work of Sutcliffe, Baylor and Menzel (1963); Riley (1963); Riley, Wangersky and Van Hemert (1964) and many other investigators has drawn attention to the formation of organic aggregates from dissolved organic matter. Earlier investigations were concerned particularly with the effect of bubbling in the formation of organic aggregates. In the near surface waters, where with regular foaming action extensive formation of aggregates occurred (e.g. Long Island Sound) such aggregates formed around the bubble surface, then adsorbed bacteria, and frequently phytoplankton cells also joined to form a mass of particulate matter of multi-origin. Such aggregates are almost certainly consumed by zooplankton and indeed by other organisms, and thus can form another pathway from dissolved organic matter to higher levels in the marine ecosystem. The formation of leptopel and its concentration into surface slicks is somewhat comparable.

But organic aggregates appear to occur at all depths in the oceans. Riley, Van Hemert and Wangersky (1965), for example, have examined the water of the Sargasso Sea, and although below ca. 500 m the amount of organic aggregates falls considerably with depth, such aggregates are present at all levels. There is clear proof that particulate matter can be formed well away from surface bubbling, and that deep water layers form particulate matter from dissolved organic substances by adsorption onto surfaces. The observations of Pomeroy and Johannes (1968) also confirm organic aggregates in deeper waters. They suggest that possibly dissolved organic matter concentrates at such surfaces, and protista, and bacteria, can then profit more succesfully on the material. Some aggregates at considerable depths may include what has been described as "marine snow" which has been observed directly in deep dives from bathyscaphes.

The aggregates will be constantly grazed upon by zooplankton, protista and other organisms, and the quantity present is the difference between their production and their depletion by grazing. The amount at any one time, therefore, may be relatively small, and it also seems likely that the rate of turnover is very low (cf. bacteria). Riley and his associates (1965) point out that as there is now general agreement regarding a relatively constant amount of particulate carbon below some 500 metres, this is not consistent with the theory of the deeper living zooplankton living regularly on a rain of particles from above. Otherwise the amount of particulate matter should decrease more or less regularly with depth. They believe that bacteria and filter feeders in the deeper zones utilize particulate matter; the amount, however, is very small and the rate of turnover excessively low. Even then, however, the amount of particulate matter should still decrease with depth unless the augmentation by aggregates by adsorption of dissolved matter onto particles is constantly operating.

Investigations such as those of Fogg (1963; 1966), Hellebust (1965) and others have clearly demonstrated that phytoplankton secretes a considerable range of substrates; carbohydrates, organic acids, peptides, aminoacids and other organic molecules of relatively small molecular weight are excreted. It is also now conceded that such materials may be excreted at any stage of cell growth. Indeed a recent interesting observation in our own laboratory from Smith and Williams (unpublished) has shown that a very significant quantity of such materials can be secreted in the very earliest phases of phytoplankton growth. Different species vary very greatly both in the rate at which they produce substances and also in the specific variety of compounds

6*

which are produced (Hellebust, 1965). This variety of relatively labile material undoubtedly contributes to the substances which are easily assimilable by heterotrophic organisms in the euphotic zone, and some small portion may be transported to deeper levels. The production of such extra cellular substances, though certainly a part of primary production, begins another pathway by which organisms, particularly heterotrophic organisms, can contribute to a food chain leading to higher levels in the ecosystem, though some of the secreted materials may be lost at least temporarily to the system.

This review of possible alternative trophic pathways in the oceans has attempted to draw most of its information, admittedly very sparse, from aphotic zones, where obviously primary production cannot proceed. Any organic material present in any form whatsoever in such deeper waters, however, must have been derived ultimately from primary production at some time and place. The alternative pathways will nevertheless be of significance in shallow seas also. In particular they will contribute to secondary production and to complex food chains, especially under conditions where photosynthesis is limited in space (e.g. to very shallow layers due to turbulence, pollution, etc.) or in time (e.g. seasonal effects in mid and high latitudes). Such additional trophic pathways must, therefore, be considered in the fertility of the seas.

References

ADAMS, J. A. and STEELE, J. H. (1966). Shipboard experiments on the feeding of *Calanus finmarchicus* (Gunnerus) *in Some Contemporary Studies in Marine Science*. (Ed. H. Barnes), 19–35.

BANSE, K. (1964). On the vertical distribution of zooplankton in the sea. *Progress in Oceanography*, 2, 53–125.

BERNARD, F. (1948). Recherches sur le cycle du *Coccolithus fragilis* Lohm. Flagellé dominant des mers chaudes. *J. du Conseil*, 15, 177–188.

BERNARD, F. (1953). Role des Flagellés calcaires dans la fertilité et la sédimentation en mer profonde. *Deep-Sea Res.*, 1, 34–46.

BERNARD, F. (1959). Elementary fertility in the Mediterranean from 0–1000 metres compared with the Indian Ocean and the Atlantic off Senegal. *Proc. Int. Oceanogr. Congr.*, 830–832. A.A.A.S. Washington.

BERNARD, F. (1963). Density of flagellates and Myxophyceae in the heterotrophic layers related to environment. Chapter 22 in *Marine Microbiology* (Ed. C. H. Oppenheimer) Springfield, Ill., Thomas. 215–228.

BERNARD, F. (1967). Research on phytoplankton and pelagic Protozoa in the Mediterranean Sea from 1953–1966. *Oceanogr. Mar. Biol. Ann. Rev.*, 5, 205–229.

DAVIS, H. C. (1953). On food and feeding of larvae of the American oyster, *C. virginica*. *Biol. Bull.* Woods Hole, **104**, 334–350.

DUURSMA, E. K. (1963). The production of dissolved organic matter in the sea as related to the primary gross production of organic matter. *Netherland J. Sea Res.*, **12**, 85–94.

DUURSMA, E. K. (1965). The dissolved organic constituents of sea water. Chapter 11 in *Chemical Oceanography*, **1**, 433–475 (Ed. Riley and Skirrow) Academic Press, London and New York.

FOGG, G. E. (1963). The role of algae in organic production in aquatic environments. *Brit. Phycol. Bull.* **2**, 195–205.

FOGG, G. E. (1966). The extracellular products of algae. *Oceanogr. Mar. Biol. Ann. Rev.*, **4**, 195–212.

FOURNIER, R. O. (1966). North Atlantic deep-sea fertility. *Science*, **153**, 1250–1252.

FULLER, J. L. and CLARKE, G. L. (1963). Further experiments on the feeding of *Calanus finmarchicus Biol. Bull.* Woods Hole, **70**, 308–320.

GOLD, K.

GUILLARD, R. R. L. (1963). Organic sources of nitrogen for marine centric diatoms. In *Symposium on Marine Microbiology* Chapter 9, 93–104. (Ed. C. H. Oppenheimer). Springfield, Ill., Thomas.

HAGMEIER, E. (1964). Zum Gehalt an Seston und Plankton im Indischen Ozean zwischen Australien und Indonesien. *Kieler Meeresforsch.*, **20**, 12–17.

HAMILTON, R. D., HOLM-HANSEN, O. and STRICKLAND, J. D. H. (1968). Notes on the occurrence of living microscopic organisms in deep water. *Deep-Sea Res.*, **15**, 651–656.

HELLEBUST, J. A. (1965). Excretion of some organic compounds by marine phytoplankton. *Limnol. Oceanogr.*, **10**, 192–206.

HELLEBUST, J. A. and GUILLARD, R. R. L. (1967). Uptake specificty for organic substrate of the marine diatom *Melosira nummuloides*. *J. Phycol.*, **3**, 132–136.

HOLM-HANSEN, O. and BOOTH, C. R. (1966). The measurement of adenosine triphosphate in the oceans and its ecological significance. *Limnol. Oceanogr.*, **11**, 510–519.

JANNASCH, H. W. (1967). Enrichments of aquatic bacteria in continuous culture. *Archiv für Mikrobiologie*, **59**, 165–173.

JOHANNES, R. E. and SATOMI, M. (1966). Composition and nutritive value of faecal pellets of a marine crustacean. *Limnol. Oceanogr.*, **11**, 191–197.

KHAILOV, K. M. (1965). Dynamic marine biochemistry—development prospects. *Oceanology*, **5**, 1, 1–9.

KHAILOV, K. M. and FINENKO, Z. Z. (1968). Interaction of detritus with high molecular weight components of dissolved organic matter in sea water. *Oceanology*, **8**, 6, 776–785.

KREY, J. (1961). Der Detritus im Meere. *J. du Conseil*, **26**, 263–280.

LEWIN, J. C. (1963). Heterotrophy in marine diatoms. In *Marine Microbiology* (Ed. C. H. Oppenheimer), Chapter 23, 220–235. Springfield, Ill. Thomas.

LIGHTHART, B. (1969). Planktonic and benthic bacterivorous Protozoa at eleven stations in Puget Sound and adjacent Pacific Ocean. *J. Fish. Res. Bd. Canada*, **26**, 299–304.

LOOSANOFF, V. L. and DAVIS, H. C. (1963). Rearing of bivalve moullusks. *Advances Mar. Biol.*, **1**, 1–136.

MCINTYRE, A. D., MUNRO, A. L. S. and STEELE, J. H. (1968). Energy flow in a sand ecosystem. Manuscript. *Symposium on Marine Food Chains*, University of Aarhus, Denmark.

MENZEL, D. W. (1967). Particulate organic carbon in the deep sea. *Deep-Sea Res.*, **14**, 229–238.

MENZEL, D. W. and GOERING, J. J. (1966). The distribution of organic detritus in the ocean. *Limnol. Oceanogr.*, **11**, 333–337.

MENZEL, D. W. and RYTHER, J. H. (1968). Organic carbon and the oxygen minimum in the South Atlantic Ocean. *Deep-Sea Res.*, **15**, 327–337.

MURRAY, J. W. (1963). Ecological experiments on Foraminifera. *J. mar. Biol. Ass. U. K.*, **43**, 621–642.

NEMOTO, T. (1968). Chlorophyll pigments in the stomach of Euphausids. *J. Oceanogr. Soc. Japan*, **5**, 253–260.

NEWELL, B. (1965). The role of detritus in the nutrition of two marine deposit feeders, the Prosobranch *Hydrobia ulva* and the bivavle *Macoma balthica*. *Proc. Zool. Soc. London*, **144**, 1, 25–45.

NISHIZAWA, S. (1966). Suspended material in the Sea: from detritus to symbiotic micro-cosmos. *Information Bulletin Planktonology Japan*, **13**, 1–33.

PINTNER, I. J. and PROVASOLI, L. (1963). Nutritional characteristics of some Chrysomonads. In *Marine Microbiology* (Ed. C. H. Oppenheimer) Chapter 11, 114–121. Springfield, Ill., Thomas.

POMEROY, L. R. and JOHANNES, R. E. (1968). Occurrence and respiration of ultraplankton in the upper 500 metres of the ocean. *Deep-Sea Res.*, **15**, 381–391.

PROVASOLI, L. and McLAUGHLIN, J. J. A. (1963). Limited heterotrophy of some photo-synthetic dinoflagellates. In *Marine Microbiology* (Ed. C. H. Oppenheimer) Chapter **10**, 105–113. Springfield, Ill., Thomas.

RILEY, G. A. (1963). Organic aggregates in sea water and the dynamics of their formation and utilization. *Limnol. Oceanogr.*, **8**, 372–381.

RILEY, G. A., VAN HEMERT, D. and WANGERSKY (1965). Organic aggregates in surface and deep waters of the Sargasso Sea. *Limnol. Oceanogr.*, **10**, 354–363.

RILEY, G. A., WANGERSKY, P. J. and VAN HEMERT, D. (1964). Organic aggregates in tropical and subtropical surface waters of the North Atlantic. *Limnol. Oceanogr.*, **9**, 546–550.

SEKI, H. (1966). Studies on microbial participation to food cycle in the sea. III. Trial cultivation of brine shrimp to adult in a chemostat. *J. Oceanogr. Soc. Japan*, **22**, 27–32.

SKOPINTSEV, B. A. (1965). Some considerations on the distribution and state of organic matter in ocean water. *Oceanology*, **6**, 3, 361–368.

SOROKIN, J. T. (1964). On the primary production and bacterial activities in the Black Sea. *J. du Conseil*, **29**, 41–60.

STEEMANN-NEILSEN, E. (1962). Contribution to Discussion; Symposium on Zooplankton Production. (Ed. J. H. Fraser and J. Corlett) *Rapp. Proc. Verb. Cons. Perm. Int. Explor. Mer*, **153**, pp. 47 and 106–107.

STEPHENS, G. C. (1967). Dissolved organic material as a nutritional source for marine and estuarine invertebrates. In *Estuaries*. (Ed. George H. Lauff) A.A.A.S., Washington, D.C.

STRICKLAND, J. D. H. (1965). Production of organic matter in the primary stages of the marine food Chain. In *Chemical Oceanography*. (Ed. Riley and Skirrow) Chapter 12, **1**, 477–610. Academic Press, London and New York.

SUTCLIFFE, W. H., BAYLOR, E. R. and MENZEL, D. W. (1963). Sea surface chemistry and Langmuir circulation. *Deep-Sea Res.*, **10**, 233–243.

SZEKIELDA, K. H. (1968). The transport of organic matter by the bottom water of the ocean. *Sarsia*, **34**, 243–252. 2nd European Symposium on Marine Biology.

TAKAHASHI, M. and ICHIMURA, S. (1968). Vertical distribution and organic matter production of photosynthetic sulfur bacteria in Japanese lakes. *Limnol. Oceanogr.*, **13**, 644–655.

VINOGRADOV, M. E. (1962). Feeding of deep sea zooplankton. *Rapp. Proc. Verb. Cons. Perm. Int. Explor. Mer*, **153**, 114–119.

WALNE, P. R. (1958). The importance of bacteria in laboratory experiments on rearing the larvae of *Ostrea edulis* (L.). *J. mar. Biol. Ass. U. K.*, **37**, 415–425.

WHEELER, E. J. (1967). Copepod detritus in the deep sea. *Limnol. Oceanogr.*, **12**, 697–702.

WILLIAMS, P. M., OESCHGER, H., and KINNEY, P. (1969). Natural radiocarbon activity of the dissolved organic carbon in the North East Pacific Ocean. *Nature*, **225**, 256–258.

WOOD, E. J. F. (1959). Some aspects of marine microbiology *J. mar. biol. Ass. India*, **1**, 1, 26–32.

WOOD, E. J. F. (1963). Heterotrophic micro-organisms in the oceans. *Oceanogr. Mar. Biol. Ann. Rev.* **1**, 197–222.

WRIGHT, R. T. and HOBBIE, J. E. (1965). The uptake of organic solutes by planktonic bacteria and algae. *Ocean Sciences and Ocean Engineering*, **1**, 116–127.

ZAIKA, V. Y. and AVERINA, T. Y. (1968). Proportions of Infusoria in the plankton of Sevastopol Bay, Black Sea. *Oceanology*, **8**, 6, 843–845.

ZENKEVITCH, L. A. and BIRSTEIN, J. A. (1956). Studies of the deep water fauna and related problems. *Deep-Sea Res.*, **4**, 54–66.

Fertilizing the sea by pumping nutrient-rich deep water to the surface*

O. A. ROELS, R. D. GERARD and A. W. H. BÉ

Lamont-Doherty Geological Observatory,
Columbia University,
Palisades, New York

Abstract

Estimates of the potential fertility of the sea can best be made in areas of natural upwelling, such as the Peru Current, where measurements of 11.2 grams carbon fixed/m²/day (Ryther, *et al.*, in press) have recently been made. This would amount to 4,090 grams carbon/m²/year, corresponding to 100 tons (dry weight) of plant material per hectare per year if the upwelling and the illumination were sustained throughout the year at the same rate, and assuming that the average carbon content of phytoplankton is 40% of dry weight (Strickland, 1965). Such a crop would require 5 tons of nitrogen per year.

Since carbon is not a limiting factor, but nitrogen frequently is, it is possible to pump deep water, rich in nitrates and phosphates, from the nutrient maximum in the sea into ponds or lagoons on shore in regions where steep offshore slopes provide access to deep water within a few miles of the coast. This situation occurs in certain areas along the coasts of Mexico, Panama, Colombia, Brazil, Chile, Peru, and Ecuador, in the Caribbean area and in relatively few areas of Africa, Asia and Australia. To bring up 5 tons of nitrogen/year would require pumping 12.2 m³ deep water per minute. Since the deep water is also cold (5°C), the cost of pumping could be offset in part by the production of fresh water, air-conditioning and electrical power by the Claude process (Gerard and

---

* Lamont-Doherty Geological Observatory Contribution # 1659.

Worzel, 1967). In the open sea, the deep water could be a by-product of deep-sea mining operations. 100 tons dry weight of plankton/hectare/year could be harvested directly or could produce 10 tons dry weight or 50 tons fresh weight of a secondary producer.

We have a pilot scheme under way on the north shore of St. Croix (U.S. Virgin Islands), where 1000 m deep water is less than one mile offshore. This will provide us with basic information on upwelling and practical information on aquaculture.

Resumen

Estimativas da fertilidade potencial do Oceano podem ser feitas melhor em áreas de ressurgência tais como as da corrente do Perú, onde foram feitas recentemente medidas de 11,2 g carbône fixado/m²/dia. Êste valor corresponderia a 4,090 grammas carbôno/m²/ano, ou 100 toneladas (pêso sêco) de material de plantas por hectare por ano, se a ressurgência e iluminação fossem mantidas a mesma razão, durante todo o ano, e assumindo que o conteúdo médio de carbôno do fitoplâncton é 40% do pêso sêco. Essa quantidade de material requereria 5 toneladas de nitrogênio/ano. Desde que carbôno, não é um fator limitante mas o nitrogênio frequentemente é, é possível bombar água de profundidade, rica em nitratos e fosfatos, do máximo de nutrientes, no Oceano, para lagoas costeiras onde o acesso de águas profundas a superfície é mais fácil. Essa situação ocorre em certas áreas ao longo das costas do México, Panamá, Colômbia, Brasil, Chile, Perú e Equador, no Mar das Caraíbas e em algumas áreas da África, Ásia e Australia. Para trazer a superfície 5 toneladas de nitrogênio/ano, dever-se-ia bombar, 12,2 m³ de água de profundidade por minuto. Desde que a água de profundidade é também fria (5°C) o custo do bombeamento poderia ser coberto, em parte, pela produção de água doce, ar condicionado e força elétrica, pelo processo de Claude. No mar aberto, a água de profundidade poderia ser um produto secundário de operações de mineração. Cem toneladas de pêso sêco de plâncton/hectare/ano, poderiam ser colhidas diretamente ou poderiam produzir 10 toneladas de pêso cêco ou 50 toneladas de pêso úmido como produto secundário.

Existe um esquema piloto na praia norte de St. Croix (Ilhas Virginicas, U.S.A.) onde a água de 1.000 metros de profundidade está localizada a menos de uma milha da praia. Êste esquema dará informações básicas sôbre a ressurgência e práticas sôbre a aquicultura.

INTRODUCTION

Just as mankind in its history on land has gone through a hunting stage and has then slowly turned to agriculture for its food production by domestication of animals instead of hunting wild ones, and by cultivating crops

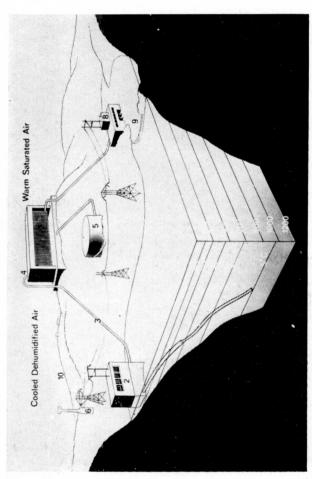

Figure 1 Cold, deep offshore water is pumped up through a large diameter pipeline and used as a coolant in a fossil-fuel or nuclear generating plant (2) and as a coolant in air-conditioning systems for a hotel complex (5). Fresh water, as a by-product of the air-conditioning system is stored in a reservoir (6). The nutrient-rich deep water, unchanged except for temperature, after use as a coolant for the power plant, is piped to artificial ponds (3) for the growth of phytoplankton. The phytoplankton bloom can then sustain organisms higher up the trophic chain. Final stripping of the nutrients from the water prior to its being returned to the sea is achieved in a lagoon (4) by the growth of commercially valuable seaweeds.

rather than picking food and berries growing wild in the forests, the day is here when we have started to plan marine harvests in a rational way.

The culture of mussels is now a large industry in Spain. Oyster culture has been introduced with varying degrees of success in different parts of the world. Shrimp farming is here, and fish are raised in many places now. The selection of fish breeding stock is a well established science.

However, to grow mussels, shrimp or marine fish, these animals must be provided with adequate food, usually marine algae. An early attempt to grow phytoplankton was made by the Pfizer Company in the New York area: Thacker and Babcock (1957) built a pilot plant for the mass culture of algae, which gave trouble-free operation over a period of almost a year. Algal growth rates of 14.4 grams of dry *Chlorella pyrenoidosa* per square meter per day were achieved in this plant. Essentially, a nutrient medium was pumped through a series of well-illuminated glass tubes, a cooling system was incorporated to remove the heat of the lamps, the media were recycled to the growth chamber, carbon dioxide-air mixture was continuously mixed into the circulating nutrient medium. The authors concluded that sunlight would be the only practical energy source, since the consumption of electrical energy, principally for illumination, brought the price of a pound of dry algae to fifty cents.

We thought, therefore, that the tropics would be a good place to stimulate primary production, and prime the marine food chain, in view of the abundant sunshine the year-round. However, the problem of supplying the mineral nutrients and the micronutrients necessary for the initiation of primary production still remained to be solved. Two scientists in our laboratories, Gerard and Worzel (1967), then came up with an idea which would supply the necessary nutrients for the photosynthetic process in the sea. They proposed a system to simultaneously produce fresh water and innovate techniques for mariculture by supplying the necessary nutrients. Figure 1 gives a schematic outline of such a plant.

Cold deep offshore water will be pumped up through large diameter pipes through a condenser area, located on shore to intercept the flow of moisture-saturated winds. This air, when cooled, condenses much of its moisture, which is conducted to storage tanks for use as drinking water. Such a scheme is obviously economically feasible only where there is deep water close to shore areas where humid winds prevail. The nutrient-rich deep water is then run into lagoons, tanks or ponds where it will stimulate the photosynthetic process by providing the necessary nutrients. The resulting

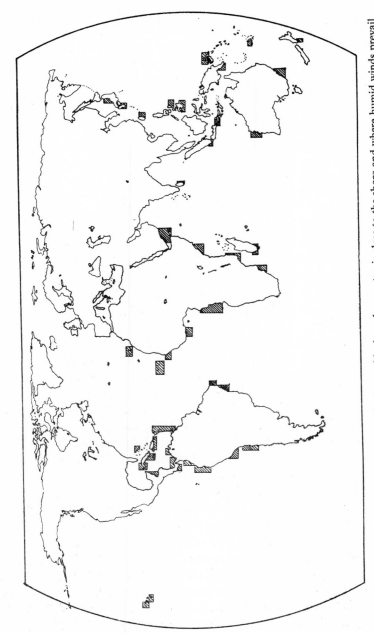

Figure 2   An outline of the various areas in the world where deep water is close to the shore and where humid winds prevail

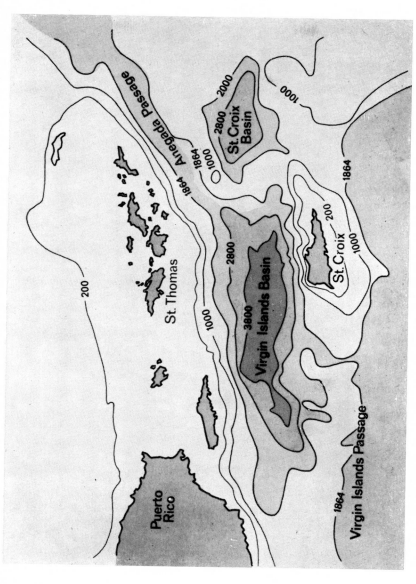

Figure 3 Illustrates why we have chosen the northern shore of St. Croix to construct our pilot plant: deep, cold, nutrient-rich water is very close to shore.

phytoplankton bloom can then sustain organisms higher up the trophic chain. Figure 2 gives an outline of the various areas in the world where these conditions are fulfilled.

Figure 3 illustrates why we have chosen the northern shore of St. Croix to construct our pilot plant: deep, cold and rich water is very close inshore there, and the moisture-laden trade winds blow constantly from the east.

THE EFFECT OF SEAWATER ENRICHMENT ON PHYTOPLANKTON GROWTH

Pinchot (1966) has suggested chemical fertilization of circular atoll lagoons by pumping deep nutrient-rich water into them to grow phytoplankton and zooplankton to culture captive whales. This proposal, which has been termed the "coral corral" would take advantage of the baleen whale's high efficiency for converting zooplankton into usable protein.

A similar project, the "Projeto Cabo Frio", has been outlined in May, 1969 by Paulo de Castro Moreira da Silva of the Instituto des Pesquisas da Marinha in Rio de Janeiro (1969).

Specific studies on nutrients limiting the production of phytoplankton in the Sargasso Sea, with special reference to iron, were done by Menzel and Ryther (1961). These authors found that surface water samples from the Sargasso Sea, with nitrate, phosphate, and vitamins added separately, or in combination, did not stimulate carbon-14 assimilation, whereas the addition of a trace-metal mixture increased carbon-14 uptake several-fold. The effective component of the trace-metal mixture was found to be iron, which alone enhanced the carbon-14 uptake for 24 hours. Addition of nitrogen and phosphorus was required, however, to produce a comparable effect over a three-day period. This work clearly indicates that after 24 hours, nitrogen and phosphorus became limiting factors. The authors believe that their conclusion was applicable to the semi-tropical and tropical western Atlantic. This work was continued by Barber and Ryther (1969), who found that upwelling out of the surface of the Cromwell Current in the eastern Pacific, was high in inorganic nutrients, relatively low in dissolved organic carbon and supported less phytoplankton growth than water to the north or south. Enrichment experiments with nitrate, phosphate, silicate, a vitamin mixture, and iron in a trace-metal mixture, showed that only the addition of a strong chelator or an undefined zooplankton extract could improve phytoplankton growth in the nutrient-rich, newly-upwelled water. The authors therefore believe that natural organic chelators, released by organisms as the water ages at the surface, may be partly responsible for the increased phytoplankton growth north and south of the equator, near the Cromwell Current in the eastern Pacific. Similarly, Tranter and Newell (1963) did enrichment experiments in the Indian Ocean and found iron more important than phosphate and nitrate to stimulate $C^{14}O_3^{--}$ uptake by phytoplankton, but their incubations were limited to 20 hours.

In contrast, Thomas (1969) showed that nitrogen was the most likely limiting nutrient for natural phytoplankton populations in the eastern equatorial Pacific Ocean, as well as off the coast of Baja California.

The tropical oceans, with their potential large productivity because of continuous and good light intensity, produce very little because of limited nutrient supply. Ryther and Guillard (1959) managed to increased CO_2-fixation by phytoplankton in tropical water by adding trace metals, especially silicate. Several workers (12, 13, 14, 15) added commercial fertilizer to Long Island Sound water and thereby greatly increased the growth of phytoplankton. This was done to grow phytoplankton in tanks at the Milford Marine Station in Connecticut to rear shellfish larvae.

It would appear as though there could hardly be a better fertilizer than nutrient-rich deep water which is responsible for the rich marine productivity in areas of natural upwelling.

RESULTS OBTAINED AT OUR ST. CROIX EXPERIMENTAL STATION

One mile due north of our beach installation on the north shore of St. Croix, the water is almost exactly 1000 m deep. Since July 1969, we have made chemical, hydrographic, and biological observations at that location and the results of a typical station (August, 1969) are shown in Figure 4. The nutrient maximum was at 800 m depth at that time. In September, 1969, the nutrient maximum had shifted slightly upward (700 m) and on October 31, 1969, the nitrate and phosphate maxima were at 600 m depth. A nitrite peak is usually observed 200–300 m above the nitrate maximum. To determine the effect of nutrient-rich water on productivity measured by carbon-14 uptake, plankton net tows were made in a rich near-shore area ("Pull Point") and the phytoplankton thus obtained was dispersed in a known volume of Millipore-filtered surface water. Aliquots of this phytoplankton dispersion, in which *Licmorphora grandis* and *Navicula* sp. were dominant, were then added to Millipore-filtered water taken during our August station at the surface and to Millipore-filtered aliquots of water taken at the same station at the same time at −800 m, where the nutrient maximum occurred (see Fig. 4). The results of this experiment are illustrated in

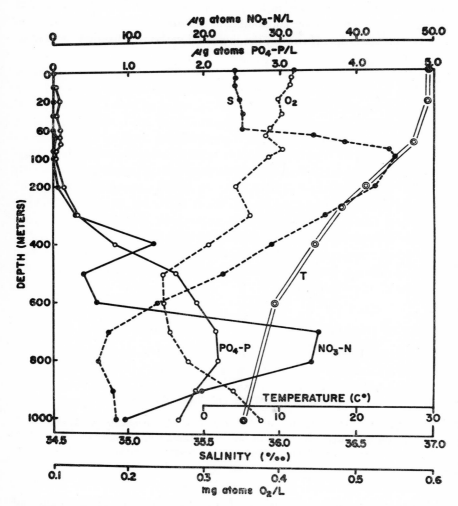

Figure 4 Chemical and physical parameters in the water column at our St. Croix station in August, 1969. This station is occupied at least once a month, and on October 31, 1969, the nutrient maximum had shifted to −600 meters

Figure 5. A 27-fold increase in carbon-14 fixation by this mixed phyto-plankton population was observed when −800 m water was used compared to the surface water taken at the same station at the same time. Although the productivity indicated was quite low, this may well have been due to the relatively poor condition of the phytoplankton after it had been caught in the net, sieved to remove zooplankters, and dispersed again in Millipore-filtered surface water. However, the increased carbon-14 fixation measured was unmistakable.

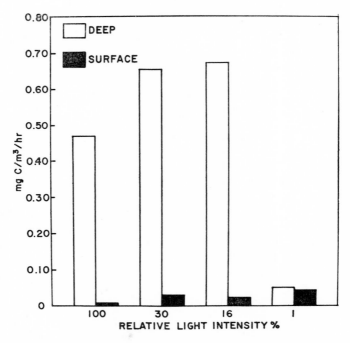

Figure 5 C^{14} fixation by mixed phytoplankton (*Licmorphora grandis* and *Navicula* sp. were dominant) in Millipore-filtered surface water on Milli-pore-filtered "deep" water from −800 m where the nutrient maximum occurred

Work now in progress at our St. Croix laboratory has clearly indicated that when *Skeletonema costatum*, which is a so-called "vitamin B-requirer", but also a typical phytoplankter thriving in natural upwelling areas, was cultivated in our laboratory with water from 0, −50, −100, −200, −400,

−600, −700, −800, −900, −1000 meters which had been Millipore-filtered, the maximum bloom of this organism occurred in water taken from −100 m depth. It was at this depth that the nutrient maximum occurred when we took the water samples on October 31, 1969 and deter-

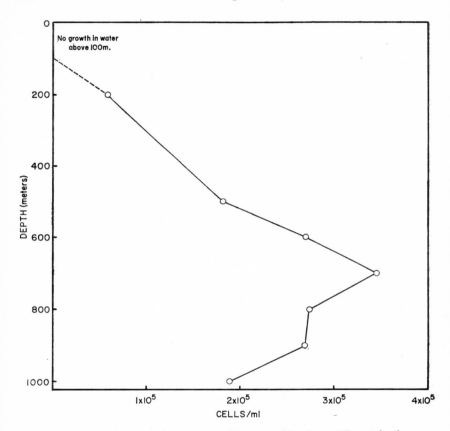

Figure 6 *Skeletonema costatum* growth in water taken from different depths

mined pH, salinity, alkalinity, nitrate, nitrite, phosphate, dissolved oxygen, and temperature at our station exactly 1 mile offshore from our beach installation on the north shore of St. Croix. It is obvious, therefore, that *Skeletonema costatum* which, according to many literature reports, requires B vitamins, thrives in nutrient-rich water taken from −600 m after it had been Millipore-filtered. The cultures were grown at 24°C. Growth of *Skele-*

tonema costatum in water from different depths is shown in Figure 6. The starting cultures contained 10^2 cells/ml. This experiment was done in our laboratory which is maintained at 24°C by air-conditioning, and with a 12 hour "light"/12 hour "dark" cycle, using fluorescent light bulbs, providing a light intensity of approximately 15% of that prevailing outside on our beach.

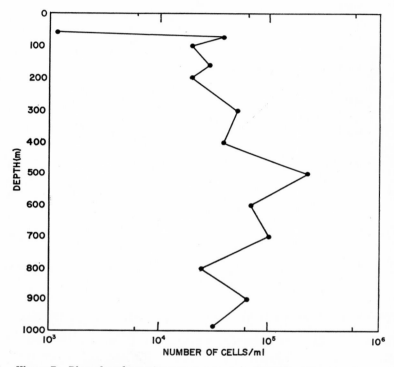

Figure 7   *Phaeodactylum tricornutum* grown in 2 weeks old water from different depths (0.45 μ Millipore-filtered) for eight days

The experiments were then repeated with *Skeletonema costatum* and with *Phaeodactylum tricornutum*. The latter does not require B-vitamins. These growth experiments were done with seawater samples which had been taken 2 weeks earlier at different depths. The temperature was 24°C and the illumination 15% of the full sunlight prevailing on the beach. The results of the experiments are shown in Figure 7 for *Phaeodactylum tricornutum* and in Figure 8 for *Skeletonema costatum*.

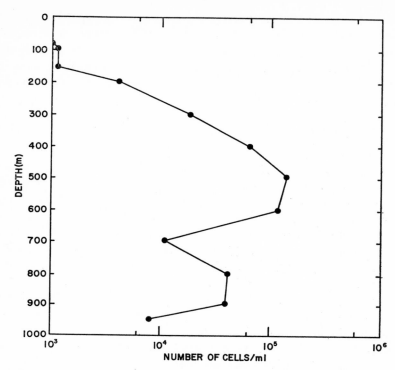

Figure 8 *Skeletonema costatum* grown in 2 weeks old water from different depths (0.45 μ Millipore-filtered) for eight days

EXPECTED MARICULTURE YIELDS AND COST OF PUMPING

We have consistently observed that at least 35 microgram-atoms nitrate nitrogen per liter seawater are present at the nutrient maximum at our St. Croix station. If we assume that this nitrate nitrogen represents approximately 65% of the nitrogen content in the water, we can then assume the total amount of nitrogen to be 54 microgram-atoms nitrogen per liter, equivalent to 756 micrograms nitrogen per liter.

We have now available on our beach property on St. Croix a 5,000 sq. ft. pond (6 ft. deep), which is equal to 464.5 sq. meters. If we use Eppley and Strickland's figure of 79 grams dry weight of phytoplankton formed per square meter per day for rich upwelling areas then we would come up with a production of 79 × 464.5 = 36.7 kilos dry weight of phytoplankton per day. This, of course, assumes that the plankton density in our 6 ft.

7*

depth will be such that most of the available solar energy will be utilized in that depth due to the high density of the phytoplankton culture. If we now divide this value by 2 for safety's sake, we would then come up with a production of 18 kilos dry weight per day, or 6,570 kilos dry weight of phytoplankton per year. This value corresponds approximately to Ryther's figure (in press) of 11.2 grams carbon fixed per m^2/day in the Peru upwelling. If we assume that plankton contains on the average 80% water, then our 6,570 kilos of dry weight plankton would correspond to 32,850 kilos of fresh weight of phytoplankton produced per year.

If we now assume a 10% efficiency of conversion of this phytoplankton to a high-priced secondary consumer, we could produce 3.3 tons of secondary producer per year in 5,000 sq. ft. To calculate the amount of nitrogen which we would have to bring up from the nutrient maximum to produce this quantity of phytoplankton, we will assume that the protein content of phytoplankton averages 37% (Parsons, Stevens and Strickland give a range of 17–57% protein for a wide variety of phytoplankters). Since we considered that we would produce 6,570 kilos dry weight of phytoplankton per year, 37% of that would be protein, or 2,431 kilos. Since the nitrogen content of protein is approximately 16%, these 2,431 kilos of phytoplankton would contain 389 kilos of nitrogen.

As we have seen earlier, the nitrogen content of our nutrient-rich deep water is 756 micrograms nitrogen/liter. This would require pumping 979 liters of deep water/minute. This amounts to 272 gallons/minute, which can be done by a very modest sized pump. Of course, as pumping rates go up, energy requirement per gallon of water pumped goes down very rapidly. The horsepower requirement for a 300 gallons/minute pump is 4 hp. and a 5 hp. motor which we have foreseen for this pump will draw 4 kilowatts. This is still a very small pumping capacity. To illustrate how horsepower and kilowatts required decrease as the capacity of the pump increases, we can cite the example of a Worthington pump, Model KL-KM wet-pit propeller pump with a capacity of 110,000 gallons/minute. This pump requires 600 hp. A pump of this capacity yields 183 gallons/minute/horsepower, versus about 26 gallons/minute/horsepower for a pump which draws 40 gallons of water per minute and requires 1-1/2 horsepower.

Moreover, the cost of pumping can be further offset by utilizing the cold temperature of the nutrient-rich water. One 60 sq. ft. cupro-bronze fin-tube coil supplied with liquid freon which could be cooled by the cold water to about 40°F by a pump supplying 300 gallons/minute of cold seawater, will produce 3,000 gallons/day of fresh water from the 77–80% humidity-laden air.

Alternatively, enough freon could be condensed to produce 55 tons or more of refrigeration, depending on the temperature differential between the deep sea water and the ambient air. Another alternative is to produce electrical power by the Claude process. The temperature differential between the ambient air and the cold sea water could conceivably produce about 12 kilowatts of electrical power when 300 gallons/minute of deep water are pumped up. The cold water could also be used for the condensers in conventional desalination plants or for cooling nuclear reactors for power production.

These aspects of the work are somewhat speculative at this time and the hard economic facts will only be learned by experience.

PRELIMINARY EXPERIMENTS ON UPWELLING IN THE OPEN SEA

Preliminary experiments will be undertaken in May, June and July, 1970, in collaboration with Deep Sea Ventures, Inc., of 12388 Warwick Boulevard, Newport News, Virginia 23606, to determine the effect of pumping nutrient-rich deep water to the surface in the open sea. Deep Sea Ventures will undertake at that time a pilot operation for deep-sea mining of manganese nodules. On board their prototype ocean mining vessel they will pump water from 2500 to 3000 ft. depth (the bottom) in the region of the Blake Plateau to bring up manganese nodules for their mining operation. We will determine the effect of dispersing this water at the surface on plankton productivity and will examine the possibility of thus establishing an "artificial upwelling" in the open sea. We are now examining the possibility of dispersing this cold and possibly turbid dense water with surface water so that it will not immediately sink through the less dense upper layers. Fluorescent dye experiments on board the ship will give a good idea of the rate of dispersion of deep water when it is returned to the ocean after extraction of the manganese nodules. We will have a variety of phytoplankton cultures ready for inoculating the area to try to determine whether we can create an artificial bloom downstream from such a mining operation. The deep water will obviously be a by-product of the mining operation and any use which can be made of it will reduce the cost of the mining operation. The initial pumping will be done through 10″ diameter pipes at a rate of 14,130 liters/minute. If we use our figure of 756 micrograms nitrogen/liter nutrient-rich water, then this pump will yield 11 grams nitrogen/minute or 16 kilograms nitrogen/day, or 6 tons nitrogen/year. This could be transformed into 93 tons of protein and the calculations can be extrapolated from that number. We will examine the possibility of establishing "marine pastures" in this fashion, downstream from a mining operation.

It should be noted that this deep-sea mining operation described here is only a pilot operation and that the planned large-scale operations would bring many times this volume to the surface.

CONCLUSION

We believe that our experiment will:

1) Enable us to study upwelling by pumping sea water from the nutrient maximum in the sea to the surface and determining its effect on marine productivity.

2) Make it possible to establish the most efficient protein production in the marine food chain using, as raw material, sea water from the nutrient maximum in the sea.

These two aspects of our project will give us a better understanding of natural upwelling and will lead to new processes for aquaculture. Moreover, this project will enable us:

1) To examine the feasibility of preparing fresh water by utilizing the cold temperature of the deep sea water to condense atmospheric moisture.

2) To examine the possibility of utilizing the cold temperature of the deep sea water for air-conditioning and other cooling purposes, e.g. in conventional desalination processes, ice-making, cooling of nuclear reactors etc.

3) To examine the feasibility of utilizing the cold temperature of the deep sea water for electrical power manufacture.

These aspects of the project will indicate whether utilization of cold, deep sea water as a new, abundant natural resource may well contribute to the development of coastal areas by providing power, fresh water, air-conditioning and marine protein.

Acknowledgements

We are grateful to Dr. E. J. Ferguson Wood and to K. Haines for their advice and help on phytoplankton at Ste Croix and to W. Hadley for the preliminary engineering calculations on pumping, fresh water production and cooling. The competent technical assistance of P. Centeno is gratefully acknowledged.

References

RYTHER, J. H., HULBURT, E. M., LORENZEN, C. J., and CORWIN, N. *The production and utilization of organic matter in the Peru coastal Current*. Texas A and M Univ. Press, College Station, Texas (in press).

STRICKLAND, J. D. (1965). In *Chemical Oceanography* Vol. I, p. 503. (Ed. Riley and Skirrow) New York: Academic Press.

GERARD, R. D. and WORZEL, J. L. (1967). Condensation of atmospheric moisture from tropical maritime air masses as a fresh water resource. *Science*, 157, 1300.

THACKER, D. R. and BABCOCK, H. (1957). The mass culture of algae. *J. Solar Energy Sci. Eng.*, 37, 50.

PINCHOT, G. B. (1966). Whale culture—a proposal. *Pr. Perspec. Biol. Med.* 10 1, 33–43.

MOREIRA DA SILVA CASTRO, PAULO DE (1969). *Projeto Cabo Frio*, Ministerio da Marinha, Rio de Janeiro, Brazil (Maio).

MENZEL, D. W. and RYTHER, J. H. (1961). Nutrients limiting the production of phytoplankton in the Sargasso sea, with special reference to iron. *Deep-Sea Res.*, 7, 276–281.

BARBER, R. T. and RYTHER, J. H. (1969). Organic Chelators: factors affecting primary production in the Cromwell Current upwelling. *J. exp. mar. Biol. Ecol.* 3, 191–199.

TRANTER, D. J. and NEWELL, B. S. (1963). Enrichment experiments in the Indian Ocean. *Deep-Sea Res.* 10, 1–9.

THOMAS, W. H. (1969). Phytoplankton nutrient enrichment experiments off Baja California and in the eastern equatorial Pacific Ocean. *J. Fish. Res. Bd. Canada*, 26, 1133 to 1145.

RYTHER, J. H. and GUILLARD, R. R. L. (1959). Enrichment experiments as a means of studying nutrients limiting to phytoplankton production. *Deep-Sea Res.* 6, 65–69.

LOOSANOFF, V. L. and DAVIS, H. C. (1963). In *Advances in Marine Biology*, Vol. I, 1. London: Academic Press.

LOOSANOFF, V. L. and ENGLE, J. B. (1942). *Science*, 95, 487.

LOOSANOFF, V. L. (1951). *Ecology*, 32, 748.

LOOSANOFF, V. L., HANKS, J. E. and GANAROS, A. E. (1957). *Science*, 125, 1092.

Some observations on the feeding of the Peruvian anchoveta Engraulis ringens J. in two regions of the Peruvian coast

BLANCA ROJAS DE MENDIOLA

Instituto del Mar
Lima, Peru

Abstract

This work deals with the feeding of anchoveta in the North, Chimbote (9° S Lat.) and in the South, Mollendo (17° S Lat.) based on 77 samples.

In the North the preference is for phytoplankton and in the South for zooplankton. These preferences are confirmed by differences in the standard-intestine length relation and in the number of gill-rakers.

Confirmation and utilization of these findings in connection with population structure problems are subjects of undergoing programs of Institute del Mar.

Resumen

Este trabajo informa sobre la alimentación de las anchovetas provenientes de la Zona Norte, Chimbote (9° L. S.) y Zona Sur de la costa peruana, San Juan (15° L. S.) y Mollendo (17° L. S.) a base de 77 muestras colectadas en 1954–1955, 1963 y 1969.

Los resultados del análisis del contenido estomacal de todas las muestras indican una marcada diferencia en los hábitos alimenticios, fitoplanctónicos para el Norte y zooplanctónicos para el Sur.

El análisis de la relaciôn longitud intestino-longitud standard y del número de branquispinas del algunas de las muestras tambien revela algunas diferencias entre ambas zonas correspondiendo el intestino más largo y el mayor únmero de branquispinas a la zona norte. La utilización de estos resultados en problemas poblacionales es materia de trabajos en marcha programados por IMARPE.

INTRODUCTION

The anchoveta *Engraulis ringens* J. is so abundant in waters off Peru that tremendous amounts have been caught annually starting with 6,800 M.T. in 1961 and increasing to 10 millions, most of which are completely used in the fish meal industry. To these figures should be added the quantities used up by the guano birds, whose main food they are, and by other species which also feed on them. This abundance must depend somehow on a reliable source of food.

As for the question as to what kinds of food this species prefers, the first studies of Vogt (1940), Sears (1941) and Rojas de Mendiola (1953, indicated a preference for phytoplankton. Later on, however, Mendiola *et al.* (1969) studied stomach contents of anchoveta caught at the same time but in 4 difference areas. Chimbote, Supe, Callao and Tambo de Mora, and found that while in Chimbote and Supe (9° S Lat. and 10° S Lat.) the contents were predominantly phytoplankton, zooplankton predominated in Tambo de Mora (14° S Lat.) and neither of them dominated in Callao.

Comparison of material from Chimbote and Mollendo also reveals the dominance of phytoplankton in the former and of zooplankton in the latter. Coincident with these findings were other studies on the average member of gill-rakers and on the length of the intestine which strongly suggest the possible existence of 2 groups of anchoveta with different alimentary habits. Such a possibility is the object of a detailed investigation program of the Instituto del Mar which will include more material systematically collected along the coast but specially at Chimbote and Mollendo.

MATERIAL AND METHODS

The present study is based on material collected in the North, Chimbote (9° S Lat.) in 1954, August to December, and in 1955, January to December as well as in the South, Mollendo (18°S Lat.), March 1954, January and April

1963, and March and June 1969, and also in an intermediate point, San Juan (15° S Lat.), April and June 1969 and finally further South, Ilo (19° S Lat.) April 1954.

Some particulars follow:

Locality	No. of samples	No. full stomachs	No. empty stomachs
Chimbote	57	402	159
San Juan	6	64	—
Mollendo	12	77	55
Ilo	2	11	5

While the stomach contents were examined in all samples, intestines were measured and gill-rakers counted only for some of the Chimbote material (Tsukayama 1965) and for all of the Mollendo and San Juan material for 1969. The samples for Chimbote were obtained from the commercial landings and the rest from experimental fishings performed during the morning hours on the assumption that at this time the commercial fishing takes place at Chimbote.

Individuals were chosen roughly randomly and put in 10% formalin to stop digestive processes, measured for standard length (68–130 mm range) and for intestine length—from pilorus to anus, the stomachs were opened and the contents examined with a binocular microscope for a preliminary appreciation of the phyto-zooplankton proportions. The contents were identified and counted, taking as a unit for zooplankton at least the head and the whole cell for phytoplankton. For both phyto and zooplankton, portions not suitable for identification because of poor condition were indicated as present (+), abundant (+ +), or very abundant (+ + +) according to visual counts. For the material mentioned previously, the gill-rakers of the first arch were counted.

The analyses of stomach contents resulted in specific tabulations only where the species were in sufficient abundance and high frequency to be considered as important as food. This was tabulated together with the number of individuals per anchoveta and the month as well as its per cent of occurrence, while other species were included as "others" within the corresponding taxonomic group. The anchovetas were classified into 2 standard length groups, larger than and smaller than 10 cm. Anchoveta with empty stomachs, however, were neither classified nor otherwise considered.

Table B shows, separated into areas, anchovetas grouped by standard length and the relation of this to intestine length.

Table C and Figure 1 deal with the gill-raker counts of our 1969 material and Tsukayama's (1965) material of 1954; we followed the latter author's methods.

From here on, we will call northern anchoveta that of Chimbote and Southern anchoveta the one from San Juan and Mollendo.

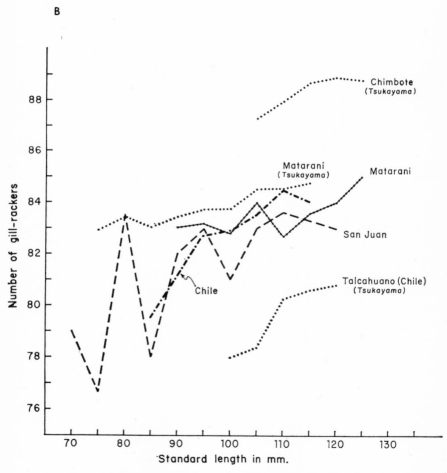

Figure 1   Comparison of the number of gill rackers with the standard length

RESULTS AND DISCUSSION

Stomach contents

Appendix Tables 1 and 2 give the principal items found in stomachs collected at Chimbote in 1954 and 1955. Diatoms were undoubtedly predominant in both years, varying sligtly according to the season.

Among those which were more important were: *Schoderella delicatula*, *Chaetoceros sp.*, *Actinocyclus octonarius* and *Nitzschia pungens*. *Skeletonema costatum*, *Coscinodiscus perforatus*, *Thalassionema subtilis* and *Thalassionema nitzchiodes* were present during almost all the year. It should be noted that some increase occurred in the number of species, and of individuals too, of dinoflagellates in the summer while copepods and euphausids decreased in numbers and percentage of occurrence.

Appendix Tables No. 3, 4 and 5 contain the names of organisms found in fish stomachs at San Juan, Mollendo and Ilo; here the number of euphausids and copepods is much larger compared with those in the North and so is their frequency. This, as well as the large size of the organisms, gives a hint about the importance of Zooplankton as food for anchovetas in the South.

Among the euphausids the most frequent were *Nyctiphanes simplex*; among the copepods, *Calanus chilensis*, *Centropages brachiatus* and different species of the genus *Oncaea*. Among the phytoplankton the only species present in almost all the samples was *Coscinodiscus perforatus*.

The predominance of zooplankton in the South in the 1953, 1963 and 1969 materials coincides with other authors observations. Schneider (1943), giving ecological data for the "chicora" *Engraulis ringens* J., says (freely translated) "its presence in our coasts (Chile) could be predicted by the detection of the type of plankton this species prefer, in which copepods abound". De Buen (1958) found remains of copepods and decapod larvae in stomachs of "chicora" collected at Iquique but the supposses that "chicora" also feeds on Phytoplankton. We have, January 1969, also examined stomachs from Arica in the Northern part of Chile, kindly sent to us by Isntituto de Fomento Pesquero de Chile, and found again that zooplankton predominates. From all this it is possible to presume that Northern Chile and Southern Peru (up to San Juan) anchoveta belong to the same group as far as alimentary habits are concerned in contrast with a Northern Peru group.

Fertility of the Sea

Comparison of the anchoveta's food in the North with that of the South

Table A, clearly shows that phytoplankton species dominate the stomach contents in the North and zooplankton ones in those of the South. This is evidenced by the numbers of diatoms, dinoflagellates, euphausids and copepods per anchoveta as well as by the percentage of occurrence of each in the total number of anchovetas.

Table A   Comparison of food of anchoveta in the North and in the South

CHIMBOTE

Organisms	1954		1955	
	Average/anch. (69 stomachs)	% occurrence	Average/anch. (333 stomachs)	% occurrence
Diatoms	315×10^4	97	226×10^4	100
Dinoflagellates	2,727	29	12,352	51
Copepoda	—	—	5	10
Euphausids	1	3	1	2

MOLLENDO

Organisms	1954		1963		1969	
	Average/anch. (26 stomachs)	% occurrence	Average/anch. (24 stomachs)	% occurrence	Average/anch. (27 stomachs)	% occurrence
Diatoms	15	46	8,193	83	901	70
Dinoflagellates	5	15	4,174	92	336	41
Copepods	10	54	60	50	171	100
Euphausids	5	54	—	—	—	—

The relation between standard length and intestine length

The relation, based on average values, is that while in the North the intestine is 1.75 times the standard length, in the South it is only 0.95 times this last measure. This large difference is better shown in Text Table 2 where the anchovetas are separated by area and classified by standard length groups.

The connection between this relation and alimentary habits in afine species is variously mentioned in the literature. For example, Harder (1958) mentions for *Engraulis ringens* J. a ratio of 1.71 : 1 between standard and intestine lengths and one of 0.85 for *Anchoa starki* G. and P, the latter feeding on zooplankton. Barrington (1957) deals with other species and states that the ones with large intestines (ratio greater than 1) are vegetarians while those with small intestines are carnivorous.

A look of Table B reveals that the northern Chile and Southern Peru material are quite alike, suggesting the carnivorous character of anchoveta there. This is in contrast with the vegetarian nature of this species in Northern Peru, a comparison which holds for all standard length groups.

Table B Relation, intestine-standard length of anchovetas from the North and from the South

Long, Std. (mm)	<100	101–110	111–120	121–130	Total of Intestine
North Area					
Chimbote	1.65	1.69	1.79	1.89	349
South Area					
San Juan (Wash-SNPI)	0.89	0.95	0.89		17
Matarani		1.30	1.24	1.23	12
Mollendo (January)		1.30	1.25		10
(March)	1.06	0.99			17
(June)	0.87	0.93	0.91	0.89	13
Average South Area	0.94	1.09	1.07	1.06	69
Chile (Arica)					
(January 1969)	0.90	0.92	0.84		51

Gill-rakers

Table C gives, by standard length groups, the average number of gill-rakers in the first arch for our material from San Juan and Mollendo and for the Tsukayama's (1966). Tsukayama (1965) found a North to South gradient from 87 for Chimbote to 83 for Mollendo. This last figure agrees with ours. The smaller number of gill-rakers in the South suggests feeding on larger organisms which could not be other than zooplankton. It should be noted that although our samples are numerically smaller than Tsukayama's (1966) the tendency of the curves is the same (Fig. 1). This figure includes the Talcahuano, Chile material of Tsukayama as well as our's North of Chile and South of Peru which again reveals great similarity among them. This is in contrast to the material from North Peru and strengths the possibility of two groups as found from stomach contents analyses as well as from standard to intestine length relation.

Table C Average numbers of gill-rackers by standard length groups in Chimbote, Callao, Pisco, Matarani e Ilo 1954 and San Juan, Mollendo 1969

Long St.	Chimbole	Callao	Pisco	San Juan	Matarani	Mollendo	Ilo
63–67				74.00			
68–72				79.00			
73–77				76.66	82.93		82.50
78–82				83.50	83.42		82.70
83–87				78.00	83.08		
88–92				82.00	83.48	83.00	
93–97				83.00	83.73	83.16	
98–102				81.00	83.82	82.83	
103–107	87.30			83.00	84.59	84.00	
108–112	87.96	88.36	85.93	83.60	84.58	82.16	
113–117	88.67	88.74	85.73		84.76	83.57	
118–122	88.82	88.13	86.16	83.00		84.00	
123–127	88.77		86.98			85.00	

CONCLUSIONS

Although, due to the scanty material, these can not be definitive, it is important to point out the following:

1. Stomach contents analyses have shown that in the North phytoplankton predominates and in the South, zooplankton predominates.

2. The standard-intestine length ratio is $1:1.75$ for the North and only $1:0.93$ for the South, although the significance of this difference is yet to be established.

3. The variation curve for the average number of gill-rakers found by us indicates a similar trend as described by Tsukayama (1965).

4. For the characteristics considered (stomach contents, standard-intestine length relation, and gill-raker counts), anchovetas from Northern Chile and Southern Peru form a group which is distinct from those of Northern Peru.

All this leads to suggest the existence of two groups of anchoveta as far as feeding habits and some anatomical characteristics are concerned; one in Northern Peru and the other in Southern Peru and Northern Chile. The presence of these two groups in central Peru can be investigated especially by means of intestine measurements, since this is a character the presumably changes slowly in spite of changes in food, and of stomach content analyses. Parenthetically, it should be noted that Southern Peru anchoveta contributes only 8 % of the total Peru landings for this species while the Mollendo landings stand for only less than 1 % of the same total (Vasquez, 1969).

In conclusion, I would like to point out that while this is only a preliminary work, there is a better collection of materials and collateral data going on, in which the collaboration of Instituto de Femento Pesquero de Chile is very much appreciated.

Acknowledgements

Thanks are expressed to Mr. Antonio Landa who read the manuscript and made valuable suggestions and to biologists Miss Noemi Ochoa and Miss Olga Gómez for their most helpful collaboration.

Table 1 Organisms present in stomachs' content

Month	August			
No. of samples	3			
Standard length in cm	<10		>10	
No. anch. with food	2		10	
No. anch. without food	14		4	
	No. cel. 10^3 per anch.	% occur-rence	No. cel. 10^3 per anch.	% occur-rence
PHYTOPLANKTON				
Diatoms centricae				
Actinocyclus octonarius			1.3	3
Chaetoceros sp.			1.5	2
Coscinodiscus perforatus			16.6	8
Roperia tessellata				
Schröderella delicatula			84.9	6
Skeletonema costatum			282.0	6
Thalassiosira subtilis			5,115.6	10
Others diatoms			37.4	70
Rest of Chaetoceros			+	
Rest of Coscinodiscus			+	
Diatoms pennatae				
Navicula sp.				
Nitzschia delicatissima				
Nitzschia pungens			79.9	80
Pleurosigma sp.			3.5	40
Thalassionema bacillaris			126.5	90
Thalassionema nizschioides			1,274.1	90
Others diatoms			4.3	30
Total diatoms			7,027.6	100
Dinoflagellates				
Ceratium furca				
Peridinium divergens				
Peridinium minutum			1.3	30
Others dinoflagellates			0.5	10
Total dinoflagellates			1.8	30
TOTAL PHYTOPLANKTON			7,029.4	100
ZOOPLANKTON*				
Eufausids				
Nyctiphanes simplex	1	100	0.1	40
Anchoveta eggs			0.1	40
TOTAL ZOOPLANKTON	1	100		

* Number of each organism

of anchovetas collected in Chimbote in 1954

September 3		October 1		November 2		December 4	
>10		>10		>10		>10	
22		6		14		15	
8		4		6		25	
No. cel. 10^3 per anch.	% occurrence	No. cel. 10^3 per anch.	% occurrence	No. cel. 10^3 per anch.	% occurrence	No. cel. 10^3 per anch.	% occurrence
0.9	9	2.9	33	62.5	86	788.9	93
6.1	23	4.5	50	2.2	21	4.7	33
12.6	64	84.4	100	18.4	50	34.1	80
0.2	4					9.4	20
12.5	18	1,125.6	100			13.8	20
6,261.8	100	451.1	100	8.1	28	118.9	33
56.7	64	96.2	67	4.5	21	55.4	27
11.6	76	38.4	100	4.0	57	13.9	67
++		++		+		+	
+		++					
		3.8	50				
				2.5	14	787.1	47
50.5	86	12.2	100	76.3	86	15.4	40
1.4	41	19.3	67	1.8	28	2.3	47
122.6	82	44.5	100			16.1	53
1,352.3	45	90.7	100				
2.7	36	7.5	67	2.5	57		
7,891.9	100	1,981.1	100	182.8	100	1,860.0	100
						3.4	27
		0.4	17			2.5	33
				1.2	21	0.7	20
		3.4	67	1.1	28	3.0	27
		3.8	67	2.3	43	9.6	47
7,891.9	100	1,984.9	100	185.1	100	1,869.6	100
0.4	54	0.1	17	0.1	14	0.1	7
0.4	54	0.1	17	0.1	14	0.1	7

Table 2 Organisms present in stomachs' content

Month	January		February		March	
No. of samples	4		4		5	
Standard length in cms.	>10		>10		<10	
No. anch. with food	30		32		1	
No. anch. without food	8		8		—	
	No. cel. 10^3 per anch.	%m occur- rence	No. cel. 10^3 per anch.	% occur- rence	N. ocel. 10^3 per anch.	% occur- rence
PHYTOPLANKTON						
Diatoms centricae						
Actinocyclus octonarius	639.2	93	123.6	84	15.1	100
Chaetoceros sp.	11.0	17	5.6	25	20.2	100
Coscinodiscus perforatus	1.7	7	0.1	3		
Rhizosolenia alata	37.8	17	0.3	3		
Rhizosolenia setigera	553.7	23	1.8	9		
Roperia tessellata	0.8	3				
Schröderella delicatula			2.3	12		
Skeletonema costatum	48.5	30	105.2	44	15.1	100
Thalassiosira subtilis	4.9	13	31.6	31	272.2	100
Others diatoms centricae	8.0	28	6.9	37		
Rest of *Chaetoceros*	++		++			
Rest of *Coscinodiscus*	++					
Diatoms pennatae						
Navicula sp.	14.7	33	1.1	6	5.0	100
Nitzschia pungens	21.2	20	8.9	41		
Pleurosigma sp.	5.8	10	0.1	3		
Pseudoeunotia doliolus	15.4	17	2.4	12	30.2	100
Thalassionema bacillaris			0.4	6		
Others diatoms pennatae			1.2	19	5.0	100
Total diatoms	1,362.8	100	291.5	97	362.8	100
Dinoflagellates						
Ceratium furca	22.2	43	4.1	34		
Peridinium divergens	1.0	7	0.1	3		
Peridinium minutum	13.1	17				
Spores of dinoflagellates	19.6	13	17.2	19		
Others dinoflagellates	15.2	20	8.1	59	5.0	100
Total dinoflagellates	71.1	57	29.5	75	5.0	100
Silicoflagellates	0.8	3	1.6	28		
TOTAL PHYTOPLANKTON	1,434.7	100	322.6	100	367.8	100

of Anchovetas collected in Chimbote in 1955

March 5		April 4				May 4	
>10		>10		<10		>10	
37		32		3		26	
12		5		4		3	
No. cel. 10³ per anch.	% occurrence	No. cel. 10³ per anch.	% occurrence	No. cel. 10³ per anch.	% occurrence	No. cel. 10³ per anch.	% occurrence
134.7	84	130.8	91	6.7	67	1.5	15
0.8	8	1.4	9	2.5	33	314.3	69
		9.4	41	18.5	67	33.9	65
		0.3	3				
8.6	24	11.3	41	5.0	33	56.9	69
15.4	22					0.2	4
11.0	32			5,365.9	67	3,163.9	46
106.8	68	14.6	16	18.5	33	7.0	8
5.0	30	28.0	56	10.9	67	19.1	50
		+		++		+++	
+				++		++	
0.4	5					1.3	23
1.4	13	13.9	59	40.3	33	62.3	85
		3.0	31			3.7	50
9.0	24	2.3	19				
2.4	11	2.4	19			1.3	15
0.5	5	0.5	6	11.8	67	18.8	73
296.0	100	217.9	100	5,480.1	100	3,684.1	100
2.6	22	1.4	9			0.2	8
0.1	3	2.8	22			0.4	4
7.1	46	3.1	25			0.4	4
4.3	30	0.6	6			0.2	4
7.9	40	6.4	53			3.8	27
22.0	76	14.3	75			5.0	38
6.8	16	3.1	16	3.4	33	2.8	19
324.8	100	235.3	100	5,483.5	100	3,691.9	100

Table 2

Month	January		February		March	
No. of samples	4		4		5	
Standard length in cms.	>10		>10		<10	
No. anch. with food	30		32		1	
No. anch. without food	8		8		—	
	No.cel.10^3 per anch.	% occurrence	No.cel.10^3 per anch.	% occurrence	No.cel.10^3 per anch.	% occurrence
ZOOPLANKTON*						
Copepods						
Calanus chilensis						
Microcalanus sp.	0.1	16	0.2	41		
Euphausids						
Nyctiphanes simplex						
Anchoveta eggs	0.1	7				
Others fishes eggs	0.1	7	0.1	6		
TOTAL						
ZOOPLANKTON	0.3	23	0.3	41		

  * Number of each organism

Month	June		July		August	
No. of samples	4		3		4	
Standard length in cms.	10		10		10	
No. anch. with food	28		25		32	
No. anch. without food	12		5		7	
	No.cel.10^3 per anch.	% occurrence	No.cel.10^3 per anch.	% occurrence	No.cel.10^3 per anch.	% occurrence
PHYTOPLANKTON						
Diatoms centricae						
Actinocyclus octonarius	3.1	43	0.2	8	0.6	19
Chaetoceros sp.	36.2	54	89.3	52	130.2	81
Coscinodiscus perforatus	2.2	43	2.3	36	8.7	69
Rhizosolenia alata	0.2	7				
Rhizosolenia setigera	123.7	36	2.5	20	0.1	6
Ropería tessellata	56.2	79	8.1	60	2.4	50
Schröderella delicatula	181.5	89	863.1	96	0.2	6

(*cont.*)

March 5		April 4		May 4			
>10		>10		<10		>10	
37		32		3		26	
12		5		4		3	
No.cel.10^3 per anch.	% occur- rence	No.cel.10^3 per anch.	% occur- rence	No.cel.10^3 per anch.	% occur- rence	No.cel.10^3 per anch.	% occur- rence
0.1	22	0.4	6				
0.1	3	0.1	12				
0.1	3						
		0.1	9			0.1	4
0.3	35	0.6	22			0.1	4

September 3		October 2		November 4		December 3	
10		10		10		10	
28		19		25		15	
2		1		16		15	
No.cel.10^3 per anch.	% occur- rence	No.cel.10^3 per anch.	% occur- rence	No.cel.10^3 per anch.	% occur- rence	No.cel.10^3 per anch.	% occur- rence
1.3	21	2.5	21	0.8	12	267.4	87
70.2	96	4.9	42	166.2	25	8.6	20
11.0	96	4.6	84	12.0	71	4.5	27
27.8	100	0.9	21	2.7	37	4.0	13
		10,298.2	95	808.0	75	0.5	7

Table 2

Month	June		July		August	
N° of samples	4		3		4	
Standard length in cms.	10		10		10	
N° anch. with food	28		25		32	
N° anch. without food	12		5		7	
	No. cel.10³ per anch.	% occur-rence	No. cel.10³ per anch.	% occur-rence	No. cel.10³ per anch.	% occur-rence
Skeletonema costatum	0.9	11	1.4	12	6.0	19
Thalassiosira subtilis	0.4	4	1.0	8	1.2	16
Others diatoms	3.0	46	16.8	64	16.8	62
Rest of *Chaetoceros*	+++		+++		+++	
Rest of *Coscinodiscus*	+				+	
Diatoms pennatae						
Navícula sp.	1.2	27	1.5	40	0.1	6
Nitzschia pungens	63.0	86	59.2	64	1.9	34
Pleurosigma sp.	6.5	61	5.7	56	3.1	34
Pseudoenotia doliolus	6.3	29	1.7	16	0.6	16
Thalassionema bacillaris	2.1	32	3.4	40	2.8	25
Others diatoms	16.6	82	213.0	84	5.7	47
Total diatoms	503.2	100	1,269.2	100	180.4	100
Dinoflagellates						
Ceratium furca						
Peridinium divergens	4.7	61	2.7	28		
Peridinium minutum	0.5	18				
Spores of dinoflagellates	0.1	4	0.6	8	0.2	9
Others dinoflagellates	2.3	36	0.9	24	1.3	34
Total dinoflagellates	7.6	61	4.2	52	1.5	37
Silicoflagellates	1.1	32	0.4	12	2.0	22
TOTAL PHYTOPLANKTON	511.9	100	1,273.8	100	183.9	100
ZOOPLANKTON*						
Copepods						
Calanus chilensis			0.3	4		
Microcalanus sp.	0.2	7	0.5	4		
Eupahusids						
Nyctiphanes simplex						
Anchoveta eggs			0.1	16	0.1	28
Other fishes eggs			0.1	8		
TOTAL ZOOPLANKTON	0.2	7	1.0	28	0.1	28

* Number of each organism

(*cont.*)

September 3		October 2		November 4		December 3	
10		10		10		10	
29		19		25		15	
2		1		16		15	
No. cel.10³ per anch.	% occurrence	No. cel.10³ per anch.	% occurrence	No. cel.10³ per anch.	% occurrence	No. cel.10³ per anch.	% occurrence
0.4	4	29.7	68	0.6	8	8.7	20
3.1	25					10.1	13
53.0	89	3,550.1	100	517.5	75	62.7	87
+++		++		+++		++	
++				++			
1.4	21					2.2	13
49.3	75	1,247.3	95	782.6	87	2.0	7
9.2	68	1.1	37				
1.9	32					26.2	47
1.5	21	3.3	26			12.1	20
3.1	43	0.1	5			4.2	20
233.3	100	15,142.7	100	2,290.4	100	413.2	100
0.3	4			0.3	12	5.0	27
3.6	50					3.0	50
3.9	50			0.3	12	8.0	50
11.4	50			0.1	4	2.0	7
248.6	100	15,142.7	100	2,290.8	100	423.2	100
				0.1	4		
				0.1	12		
				0.1	4		
0.1	14						
0.1	14			0.3	17		

Table 3 Organisms present in stomachs content of anchovetas collected

Year	1954				1963	
Month	March				January	
No. of samples	6				1	
Standard length in cms.	<10		>10		>10	
No. anch with food	5		21		11	
No. anch. without food	5		30		1	
	No. cel. 10^3 per anch.	% occurrence	No. cel. 10^3 per anch.	% occurrence	No. cel. 10^3 per anch.	% occurrence
PHYTOPLANKTON						
Diatoms centricae						
Actinocyclus octonarius					0.087	9
Biddulphia longicruris					0.698	36
Coscinodiscus perforatus	0.022	80	0.008	38	0.011	9
Rhizosolenia alata						
Roperia tessellata					0.087	9
Skeletonema costatum					0.437	18
Thalassiosira subtilis					0.785	36
Thalassiosira sp.					0.087	9
Diatoms pennatae						
Asterionella japonica						
Navicula sp.					0.262	18
Nitzschia closterium						
Pleurosigma sp.						
Thalassionema bacillaris					0.087	9
Thalassionema nitzschiodes					0.262	27
Total diatoms	0.022	80	0.008	38	2.803	82
Dinoflagellates						
Ceratium furca	0.006	20	0.005	14	0.131	36
Dinophysis acuminata					0.262	27
Dinophysis tripos					0.262	27
Peridinium pentagonum						
Peridinium peruvianum						
Prorocentrum micans					1.309	82
Others dinoflagellates						
Total dinoflagellates	0.006	20	0.005	14	1.964	91
TOTAL PHYTOPLANKTON	0.028	80	0.013	48	4.767	91

in Mollendo March 1954, January and April 1963, March and June 1969

1963		1969				
April 1		March 2			June 2	
>10 13 2	<10 15 0	>10 2 0	<10 5 5		>10 5 12	
No. cel. 10³ per anch.	% occur-rence	No. cel. 10³ per anch.	% occur-rence	No. cel. 10³ per anch.	% occur-rence	No. cel. 10³ per anch.	% occur-rence	No. cel. 10³ per anch.	% occur-rence

No. cel. 10^3 per anch.	% occur-rence	No. cel. 10^3 per anch.	% occur-rence	No. cel. 10^3 per anch.	% occur-rence	No. cel. 10^3 per anch.	% occur-rence	No. cel. 10^3 per anch.	% occur-rence
0.147	15					0.007	40	0.003	20
		0.048	7						
11.150	38								
1.033	15	0.096	7						
0.886	31	1.248	53	0.360	50				
0.147	8								
0.147	8	0.048	7						
		0.096	13						
0.073	8								
		1.526	87	0.180	50				
13.583	85	3.062	93	0.540	100	0.007	40	0.003	26
0.184	31	0.648	60						
0.221	15								
1.107	38	0.048	7						
		0.145	20						
0.147	15	0.048	7						
2.658	77								
2.067	69	0.096	13	0.360	50				
6.384	92	0.985	67	0.360	50				
19.967	100	4.047	93	0.900	100	0.007	40	0.003	20

Table 3

	1954				1963	
Year						
Month	March				January	
No. of samples	6				1	
Standard length in cms.	<10		>10		>10	
No. anch. with food	5		21		11	
No. anch. without food	5		30		1	
	No. cel. 10^3 per anch.	% occur- rence	No. cel. 10^3 per anch.	% occur- rence	No. cel. 10^3 per anch.	% occur- rence
ZOOPLANKTON*						
Copepods						
Acartia clausi					86	64
Calanus chilensis						
Candacia sp.						
Centropages brachiatus						
Corycaeus sp.						
Copepod sp.	15	100	5	43		
Euchaeta marina						
Microsetella sp.					1	9
Oncaea mediterranea					1	9
Rest of copepods					+++	
Euphausids						
Euphausia mucronata			1	9		
Nyctiphanes simplex	5	20	4	52		
Ostracods					2	45
TOTAL ZOOPLANKTON	20	100	10	81	90	64

* Number of each organism

(*cont.*)

1963		1969							
April				March		June			
1				2		2			
>10		<10		>10		<10		>1	
13		15		2		5		5	
2		0		0		5		12	
No. cel. 10^3 per anch.	% occurrence	No. cel. 10^3 per anch.	% occurrence	No. cel. 10^3 per anch.	% occurrence	No. cel. 10^3 per anch.	% occurrence	No. cel. 10^3 per anch.	% occurrence
		429	100	205	100				
								8	20
						1	40	1	40
31	31	2	47					2	20
						1	20	1	20
1	31	6	27			1	20	2	60
						11	100	7	80
		1	7					1	20
		2	33			2	60	2	20
++		+++		+++		+++		+++	
1	8								
33	38	439	100	205	100	16	100	24	100

Table 4 Organisms present in stomachs' content of anchovetas collected in San Juan, April and June 1969

Month	April		June			
No. of samples	1		5			
Standard length in cms.	>10		<10		>10	
No. anch. with food	7		32		25	
No. anch. without food	0		0		0	
	No. cel. 10³ per anch.	% occurrence	No. cel. 10³ per anch.	% occurrence	No. cel. 10³ per anch.	% occurrence
PHYTOPLANKTON						
Diatoms centricae						
Actinocyclus octonarius	0.600	29	0.091	16	0.114	36
Coscinodiscus perforatus			0.030	78	0.074	88
Chaetoceros sp.	0.180	29	0.044	6	0.211	12
Roperia tessellata	0.030	14			0.038	4
Thalassiosira sp.	0.840	14	0.384	12	0.388	12
Diatoms pennatae						
Asterionella japonica	0.120	14				
Nitzschia pungens	0.120	14				
Pleurosigma sp.	0.030	29	0.020	9	0.020	12
Thalassionema						
nitzschiodes	16.554	100	1.140	9	1.076	20
Total diatoms	18.474	100	1.709	81	1.921	96
Dinoflagellates						
Dinophysis tripos	0.020	29				
Others dinoflagellates			0.112	16	0.135	8
Total dinoflagellates	0.020	29	0.112	16	0.135	8
Silicoflagellates	0.040	29				
TOTAL PHYTO-PLANKTON	18.534	100	1.821	81	2.056	96
ZOOPLANKTON*						
Copepods						
Acartia clausi			1	6		
Calanus chilensis			13	94	18	96
Centropages brachiatus	8	100	1	22	1	16
Copepods sp.			2	37	1	4
Microcalanus sp.	4	86	2	9	1	12
Microsetella sp.	1	29	1	6	1	8
Oncaea mediterranea			1	34	1	24
Oncaea sp.	2	71	2	53	3	48

* Number of each organism

Table 4 (*cont.*)

Month	April	June	
No. of samples	1	5	

Standard length in cms.	>10	<10	>10
No. anch. with food	7	32	25
No. anch. without food	0	0	0

	No. cel. 10^3 per anch.	% occurrence	No. cel. 10^3 per anch.	% occurrence	No. cel. 10^3 per anch.	% occurrence
Euphausids						
Nyctiphanes simplex			21	81	10	60
Euphausia mucronata			1	12		
Euphausids sp.			2	22	1	8
Ostracods			1	3	1	8
TOTAL ZOOPLANKTON	15	100	48	100	38	100

Table 5 Organisms present in stomachs' content of anchovetas collected in Ilo, April 1954

Month	April	
No. of samples	2	

Standard length in cms	<10	>10
No. anch. with food	11	0
No. anch. without food	1	4

	No. cel. 10^3 per anch.	% occurrence
PHYTOPLANKTON		
Diatoms centricae		
Actinocyclus octonarius	0.157	27
Coscinodiscus perforatus	0.090	27
Thalassiosira subtilis	0.028	18
Diatoms pennatae		
Nitzschia pungens	0.148	27
Thalassionema bacillaris	0.006	9
Total diatoms	0.429	45
Dinoflagellates		
Ceratium furca	0.009	18
TOTAL PHYTOPLANKTON	0.438	45
ZOOPLANKTON*		
Copepods		
Copepods sp.	2	9

* Number of each organism

Table 5 *(cont.)*

Month	April	
No. of samples	2	
Standard length in cms	<10	>10
No. anch. with food	11	0
No. anch. without food	1	4

	No. cel. 10^3 per anch.	% occurrence
Euphausids		
Euphausia mucronata	2	45
Nyctiphanes simplex	7	73
TOTAL ZOOPLANKTON	11	100

References

BARRINGTON, E. J. W. (1957). *The Physiology of Fishes*. III Alimentary canal and digestion. New York: Academic Press Inc.

CUPP, E. E. (1943). Marine plankton diatoms of the West Coast of North America. *Bull. Scripps Ins. Ocean.* **5**, 1, 1–238.

DE BUEN, FERNANDO (1958). Peces de la Superfamilia *Cupleidae* en aguas de Chile. *Rev. de Biología Marina* Vol. VIII. Publicado por la Estación de Biologia Marina de la Universidad de Chile.

HARDER, WILHELM (1958). El intestino como caracter diagnóstico para la identificación de ciertos cupleoides (*Engraulidae, Clupeidae, Dussumeriidae*) y come caracter morfométrico para la comparación de las poblaciones de anchoveta (Cetengraulis mysticetus) *Inter-Amer. Trop. Tuna Comm. Bull.* **II**, 8.

MANN, GUILLERMO F. (1954). *La vida de los peces en aguas chilenas*. Universidad de Chile.

ROJAS, E. BLANCA (1953). Estudios preliminares del contenido estomacal de las anchovetas. *Bol. Cient. de la Cía. Adm. del Guano.* **1**, 1, 33–42.

ROJAS DE MENDIOLA, BLANCA (Unpublished). Breve informe sobre los hábitos alimenticios de la anchoveta (*Engraulis ringens* Jennyns) en los años 1954–1958. Informe presentado a la Cía. Administradora del Guano el 30 de abril de 1959.

ROJAS DE MENDIOLA, BLANCA, OCHOA, NOEMI, *et al.* (1969). Contenido estomacal de anchoveta en cuatro áreas de la costa peruana. Inf. 27. *Inst. del Mar del Perú-Callao.*

SCHNEIDER, OLIVER C. (1943). *Catálago de los peces marinos del litoral de Concepción y Arauco.* Museo de Concepción. Chile.

TSUKAYAMA, ISABEL (1966). El número de branquispinas como caracter diferencial de sub-poblaciones de anchoveta (*Engraulis ringens* J.) en las costas del Perú. I Seminario Latino-Americano sobre el Océano Pacífico Oriental—Univ. Nac. Mayor de San Marcos.

VASQUEZ, ISAAC (1969). Resumen general de la pesquería de la anchoveta durante el año 1968. Informe Especial No IM-35. *Inst. del Mar del Perú-Callao.*

Benthic biomass and surface productivity*

GILBERT T. ROWE

Woods Hole Oceanographic Institution
Department of Biology
Woods Hole, Massachusetts 02543

Abstract

Benthic samples from the north temperate Atlantic, the Gulf of Mexico, the Atlantic off Brazil, and the Pacific off Peru provided data for a comparison of animal densities and biomass under varying ecological conditions.

The relationships between the logarithm (base 10) of biomass (or animal density) and depth can be described by statistically significant least squares linear regressions. The average biomass and the different rates of decrease in life with depth in different regions can be used to infer the magnitude of effects of surface production on the bottom fauna. These averages are significantly higher in regions of high primary productivity (New England and Peru) than where productivity is low (Bermuda, Brazil, and the Gulf of Mexico). The regression coefficients or rates of decrease in animal density were greatest where surface productivity varied markedly in an offshore direction (Woods Hole and Brazil). Where productivity varied to a lesser degree (Peru, Gulf of Mexico, and Bermuda), the rates of decrease were reduced. These regressions suggest that while depth exerts the most stringent effects, surface productivity ranks second in controlling benthic biomass.

In the upwelling region off Peru, the hydrography and associated high productivity have caused marked oxygen depletion in sediments and bottom water. The result is a numerically dense fauna of low biomass and diversity where the oxygen minimum zone impinges on the bottom, and a diverse community of high biomass in deeper water offshore. In this

* Contribution No. 2420 from the Woods Hole Oceanographic Institution.

situation, the effects of productivity on benthic biomass are more pronounced than depth, and rather than ameliorating benthic production, have
inhibited it.

Resumen

Amostras de bentos do Atlântico Norte temperado, do Golfo do México,
do Atlântico ao largo das costas do Brasil, e do Pacífico ao largo do
Perú, forneceram material para comparar as densidades de biomassa e de
animais em diversas condições ecológicas.

As interrelações entre o logarítimo (base 10) da biomassa (ou densidade
animal) e profundidade podem ser descritos por regressões lineares,
estatisticamente significantes.

A biomassa média e as diferentes razões de decréscimo de vida, com a
profundidae, em diferentes regiões, podem ser usadas para inferir a magnitude dos efeitos da produção da superfície na faixa de fundo. Essas médias
são significantemente mais altas em regiões de alta produtividade primária
(Nova Inglaterra e Perú) do que em regiões onde a produtividade é baixa
(Bermuda, Brasil e Golfo do México).

Os coeficientes de regresão ou razões de decréscimo da densidade animal
são maiores onde a produtividade de superfície varia de forma marcante
(Woods Hole e Brasil), da costa ao oceano. As razões de decréscimo foram
menores onde a produtividade variou em menor escala (Perú, Golfo do
México e Bermuda).

Essas regressões sugerem que enquanto a profundidade é responsável
pelos efeitos mais marcantes, a produtividade de superfície é o segundo
fator mais importante no contrôle da biomassa bêntica. Na região de
ressurgência, no Perú, a hidrografia e a alta produtividade a ela associada,
causaram uma marcante depleção de oxigênio nos deimentos e na água do
fundo. O resultado é uma fauna numèricamente densa, de baixa biomassa
e diversidade, próxima a costa, e uma comunidade diferente de alta biomassa, em águas mais profundas. Nessa situação, os efeitos da produtividade
na biomassa bêntica são mais pronunciados do que os da profundidade e
inibiram a produção bêntica.

INTRODUCTION

It has been established empirically that animal density (numbers) and
biomass decrease with increasing depth, and distance from land (See Sanders
and Hessler, 1969, for a recent review.). It has also been strongly implied
that the quantity of animal life remains relatively constant from basin to
basin (Sanders, Hessler, and Hampson, 1965) and is more or less independent
of surface productivity (Frankenberg and Menzies, 1968). To explore these

postulates, a comparison of several different studies is presented with the hope biomass differences between basins can be defined and inferences can be made about sources of energy available to deep-sea animals.

METHODS

Few thorough quantitative studies have been made on deep-sea benthos, and sampling and sorting vagaries generally prevent valid comparisons between different publications. Because of this, concern here is with recently

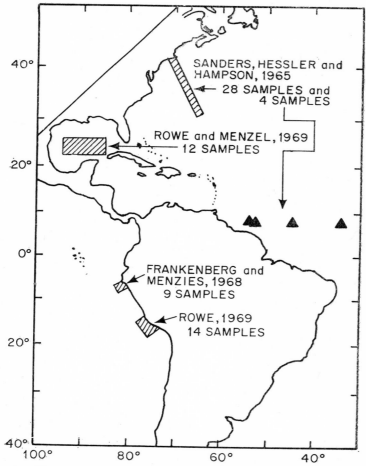

Figure 1 Western hemisphere with regions of study outlined

published data (Sanders, Hessler, and Hampson, 1965; Frankenberg and
Menzies, 1968; and Rowe and Menzel, 1969) accrued through techniques
with which I have been personally familiar and which are explicitly described
in print. (In brief, Frankenberg and Menzies used a Campbell grab and
sieved through 0.55 mm screens while the other studies used an anchor
dredge or VanVeen grab and sieved through 0.42 mm screens.)

   In the above references, data can be presented comparing biomass (in
terms of animal wet weight and/or numbers of animals) with depth for the
boreal water off New England (Sanders, Hessler, and Hampson, 1965), the
Sargasso Sea off Bermuda (Sanders, Hessler, and Hampson, 1965), the
tropical Atlantic off Brazil (Sanders, Hessler, and Hampson, 1965), the
southern Gulf of Mexico (Rowe and Menzel, 1969), and two areas of the
Peru tropical upwelling (Fig. 1) (Frankenberg and Menzies, 1968 and
14 original samples). These amount to a total of 67 samples for which animal
numbers were available. Of these, 35 were also measured in wet weights of
the animals.

   To approach the problem of what environmental variables besides depth
control biomass, the effects of depth must be eliminated from consideration,
and this has hopefully been accomplished by our statistical approach.
Defining the biomass of each region in terms of a least squares regression or
estimating equation relative to depth allows a comparison of the differences
in the regions, both in terms of the rates of decrease with depth and the
average biomass represented by the differences in displacement of the lines
above the abscissa. Analysis of variance likewise allows placing confidence
limits on these separations.

ANIMAL NUMBERS

To define a general relationship of the abundance of life with depth, a least
squares regression or estimating equation has been calculated for all animal
densities [$Y = 3.4 - 0.00037 (X - \overline{X})$ where $Y = \log_{10}$ of nos./m^2 ($P < 0.001$
b = 0)]. This provides an estimation of the number of animals per square
meter at any given depth, but under a wide range of ecological conditions.
Although the general relationship defined here has long been observed
(Sanders and Hessler, 1969), it has never been mathematically defined. Using
this as a basic model, the homogeneity between the various basins can be
tested, and from the slopes and means of the different regressions, it can be
ascertained whether or not basins differ in the numbers of animals they support.

Boreal versus tropical continental margin

Sanders, Hessler, and Hampson (1965) noted that their estimates of animal numbers per square meter off Massachusetts were far above previously published reports, and they concluded this was the result of using a finer meshed sample sieving screen and taking great care with sorting in the laboratory. To test this, they took four samples off the northeast coast of South America and concluded the difference in numbers there was not significant enough to warrant any other conclusions. These densities were converted to logarithms (base 10) and estimating equations were calculated:

$Y = 4.1 - 0.00044 (X - \bar{X}) (P < 0.001 \ b = 0)$ for Massachusetts and
$Y = 3.7 - 0.00045 (X - \bar{X}) (P < 0.05 \ b = 0)$ for Brazil (Fig. 2, Insert).

Based on analysis of variance, the rates of decrease in animal densities can be presumed equal, but means of the regressions differ significantly by a factor of $0.4 \log_{10}$ units ($P < 0.001 \ \bar{Y}$ Mass. $= \bar{Y}$ Brazil) and it can be suggested that deep-sea animal densities do differ in a statistically predictable manner. Based solely on these two areas it can be suggested also that the rates of decrease with depth are equivalent, but average densities may be significantly higher at high latitudes, depending on some as yet undefined parameters.

Offshore in the Atlantic

The Sanders transect south from New England continued to Bermuda and therefore crossed the Sargasso Sea overlying the northern border of the Hatteras Abyssal Plain, and the Bermuda Rise and Pedestal. The regression (Fig. 2, Insert) of data from this area was $Y = 2.8 - 0.00021 (X - \bar{X}) (P < 0.01 \ b = 0)$, and although the Massachusetts and Bermuda stations merged on the Abyssal Plain, both their rates of decrease ($P < 0.1$) and the mean number of animals ($P < 0.001$) differed. In other words, moving from a region of equal low density and great depth (i.e., 5000 m on the Hatteras Abyssal Plain) into shallower water, the rate of increase in life was greater moving north to the continental margin than south to Bermuda. It might now be reasonable to say the difference in biomass manifested in these regressions was the "continental" contribution, since the northern section lies adjacent to the continent and the southerly half led only to the small island of Bermuda.

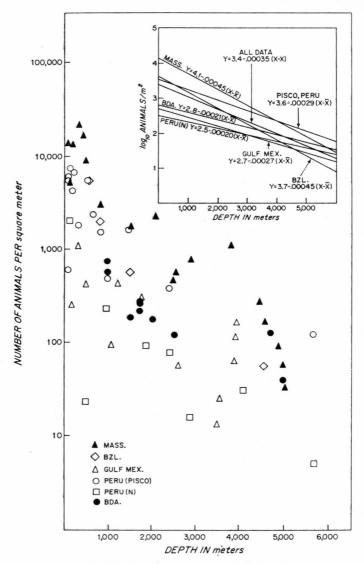

Figure 2 Estimates of numbers of animals per square meter versus depth based on the 67 samples from the regions outlined in Figure 1. Insert illustrates least squares regressions of the data from each region

The Gulf of Mexico, a basin of low productivity

The Gulf of Mexico is a relatively shallow basin, which, although supporting some economically important commercial fisheries, appeared to be characterized by a depauperate deep benthic biomass. Utilization of data from 12 samples from the southern Gulf (Rowe and Menzel, 1969) provided the following regression: $Y = 2.7 - 0.00027 (X - \bar{X}) (P < 0.01\ b = 0)$, and the conclusion can be drawn that the Gulf of Mexico supports only about as many individuals on its abyssal plain at a depth of *ca.* 3400–3700 m as does the Atlantic on the Hatteras Abyssal Plain at *ca.* 5000 m, the density being approximately 50–100 individuals/m^2 (Fig. 2, Insert).

Tropical upwelling off Peru

Data off Peru come from two sources and are somewhat contradictory. Frankenberg and Menzies (1968) collected 9 grab samples at depths of 126 to 6229 m off northern Peru; and since their density estimates were of a magnitude similar to those of Sanders *et al.*, they concluded that the deep benthos off Peru is unaffected by high surface productivity.

More recently, I had the opportunity to take 14 grab and anchor dredge samples just south of Pisco, Peru, while aboard the Woods Hole Oceanographic Institution's R.V. Gosnold and the R.V. T. G. Thompson, operated by the University of Washington. The density data taken along the northern coast (Frankenberg and Menzies, op. cit.) followed the regression $Y = 2.5 - 0.00020 (X - \bar{X}) (P < 0.05\ b = 0)$, whereas the Pisco equation was $Y = 3.6 - 0.00029 (X - \bar{X}) (P < 0.001\ b = 0)$ (Fig. 2, Insert). While the rates of decrease were only suggestively (<0.05) separable, average animal numbers per square meter in the 14 samples taken off Pisco were significantly above ($P < 0.001\ \bar{Y}$ Pisco $= \bar{Y}$ North Peru) those taken some 360–600 nautical miles to the north. Ecologically, this is difficult to explain, since presumably both areas are affected by essentially equivalent hydrographic conditions. The differences were therefore at first considered to be the anomalous result of vagaries in sample sorting techniques. Perusing the Frankenberg and Menzies table of data (1968, p. 625) provided a partial answer to the problem. No "Nematoda (large)" were counted at any of their shallow stations (126, 519, and 995 m deep), whereas off Pisco to the south, "all" nematodes above the limiting screen mesh of 0.42 mm were counted, and as a result, this group accounted for as much as 97% of the total number of animals at stations off Pisco within this depth range.

This demonstrates that the density of animals unfortunately may not always be a valid ecological idicator of biomass. Although the studies of Sanders, Hessler, and Hampson (1965) have used animal numbers alone to draw conclusions concerning biomass, it has been suggested (Rowe and Menzel, 1969) that this may be a poor estimate relative to wet weights, dry weights, and also the organic carbon in the organisms. This suggestion was based on the relative slopes of linear regressions of these different measures of biomass on depth, the decrease in numbers being suggestively slightly less than the decrease in the weight measures, which all had the same slope.

ANIMAL WET WEIGHTS

Fortunately, the samples from the Gulf of Mexico, off northern Peru and off Pisco were also weighed wet, and a comparison of this measure of biomass will perhaps provide a better measure of whether or not these areas are different (Fig. 3).

The Gulf of Mexico versus Peru (northern)

The Gulf of Mexico, a depauperate basin in terms of animal numbers, can also be represented by an estimating equation of $Y = 3.6 - 0.00054$ $(X - \bar{X})(P < 0.001\ b = 0)$, where Y is the $\log_{10}$ of wet weight in milligrams and X is depth in meters (Fig. 3, Insert). Likewise, the northern Peru equation is $Y = 4.3 - 0.00027 (X - \bar{X}) (P < 0.01\ b = 0)$. Both the slopes and the means of these regressions are significantly different ($P < 0.01$), and it can be concluded that they differ ecologically.

A more difficult question is differentiating between northern Peru and off Pisco, and complicating the question is the realization that the latter wet weight data do not follow a linear regression on depth with any confidence. As a result, statistical comparisons are difficult. Some stringent environmental parameter must be overriding the limitation that depth puts on biomass, even though animal numbers responded significantly to depth in the usual, inverse fashion.

Examination of the original data (Fig. 3) illustrates that wet weights off Pisco follow an eccentric pattern compared to other regions. At a depth of 300 m where considerable biomass is expected, estimates were less (600 mg/m²) than any other sample from less than 500 m, but at a depth of 85 m, a Pisco sample provided an estimate of 56,820 mg/m², or higher than any

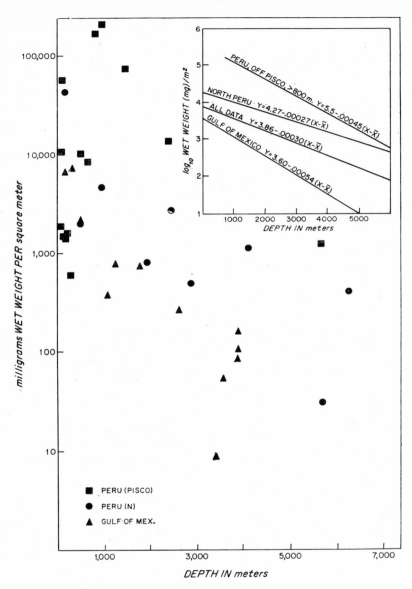

Figure 3 Estimates of wet weight of animals per square meter versus depth based on the 35 samples from the Gulf of Mexico and off Peru. Insert ilustrates least squares regressions of the data; the line for data from Pisco, Peru, considers only the samples from greater than a depth of 800 m

other sample from less than 500 m. More remarkable than this high variance in shallow water was the increase from the low at 300 m to the astonishingly high values at slightly greater depths. A maximum estimate of 204,504 mg/m^2 was made from an anchor dredge sample at 1000 m and not far behind was an estimate of 165,205 mg/m^2 from a grab sample at 875 m. At 1000 m in the Gulf of Mexico about 1000 mg/m^2 would be expected (from the estimating equation above, Fig. 3, Insert) and off northern Peru this would increase to 10,000 mg/m^2. From the maximum centered near 1000 m off Pisco, biomass dropped off predictably as depth increased. A regression based on the 5 stations from greater than 800 m followed the equation $Y = 5.5 - 0.00054\,(X - X)\,(P\,0.01\;b = 0)$. This line (Fig. 3, Insert), representing the maximum in biomass for any region, is separable from all other regions, including northern Peru.

DISCUSSION: ECOLOGICAL PARAMETERS OF BIOMASS

Depth

Our ability to define biomass relative to depth in a statistically reliable manner, irrespective of other variables in the environment, suggests that something directly related to the length of the water column limits life. Since the consensus is that benthic biomass is generally limited by how much food reaches the bottom, it can be concluded that something in the water column is removing utilizable organic energy from it.

The problem of what and how this happens will not be approached here, but despite the profound effects of depth on biomass, a search for reasonable explanations of why different basins support different amounts of life can be attempted.

Primary production

In general, the differences in average biomass for the widely separated areas appear to be related to primary production. For example, the data from between Bermuda and Massachusetts indicate biomass increases more markedly toward the continent, where annual productivity is estimated to be 120–165 gm C/m^2/year. (Ryther, 1963), than near Bermuda, where annual production is somwhat less 72 gm C/m^2/year, Menzel and Ryther, 1960). Likewise the Gulf of Mexico has relatively low productivity in its offshore waters, which correlates with its low benthic biomass. Bogdanov, Sokolov, and Khromov (1968) suggested central Gulf production was low, and estimates, based on W.H.O.I. samples (Corwin, 1969), were *ca.* 25 g C/m²/year for offshore water in the western Gulf. Brazil, falling in between the high Massachusetts and the low Gulf of Mexico biomasses, was also characterized by intermediate levels of productivity (Ryther, Menzel, and Corwin, 1967), although these latter measurements are scattered and are not sufficient for making estimates of production exactly where the benthic samples were taken. A geostrophic uptilting of nutrient rich subsurface water between the Guiana Current and the coastline, caused by the lens of fresh water debouching from the Amazon, ameliorated production nearshore to some degree, and estimates of production nearshore should be slightly above those in the tropics elsewhere. The large rate of decrease in animals ($b = 0.00045$), although probably not too reliable an estimate because of the few samples taken, might reflect the differences in the productive nearshore environment somewhat atypical for the tropics, and offshore water unaffected by Amazon outflow and therefore characterized by water of extremely low productivity (*ca.* 20 gm C/m²/year). In other words, the maximum rate of decrease in numbers of animals here may correlate with a marked gradient in surface production.

The salient correlations between productivity and biomass just described are less complex than off Peru, especially since the wet weight biomass off Pisco does not follow a linear regression. Rough estimates of production in the upwelling along the western margin of continents are *ca.* 300 gm C/m²/year (Ryther 1969), an amount far above the estimates for the basins referred to above. At all shallow stations, however, the numbers of animals off Massachusetts rank far above the Peru samples, even though production off Peru is double that off New England. The Pisco, Peru numbers, however, decrease at a slower rate (0.00029) than do those from Massachusetts (0.00044), so that at depths greater than about 4000 m, more animals can be expected in the Peru–Chile trench than at equivalent depths on the North American continental rise or the Hatteras Abyssal Plain. An explanation for this reversal may be that over the Massachusetts transect production varies from *ca.* 150 gm C/m²/year to about 60 gm C/m²/year in the Sargasso Sea, whereas the deeper water off Peru lies near shore under water whose production can still be estimated to be close to 300 gm C/m²/year. The lower numbers in shallow water off Peru may be a manifestation of the stress induced by low oxygen, as discussed below.

A more difficult and perhaps less reliable task is relating differences between Pisco and northern Peru to differences in surface production, but this warrants close attention because it may give insight into just how responsive the benthos is to spatial variations in production. Numerous measurements of primary production have been made along the coast of Peru (Ryther, 1966 and others), but due to the temporal and spatial nature of the measurements, I have found no statistically reliable data verifying that production is indeed greater off Pisco over the depths in question. The variance in the measurements is too great to make this a reliable generalization.

It may be, however, that upwelling, the cause of the high productivity off Peru, is more persistent off Pisco than elsewhere, and this may be the reason for the biomass differences in the regions. Reference to the monthly sea surface temperature maps published by the Tuna Resources Laboratory of the U.S. Bureau of Commercial Fisheries suggests that the water just south of Pisco was several degrees colder than surrounding water for every month for the last three years, at least during those months when records were sufficient for comparison. To the north no such persistence of cold water was noted. This can be suggested as an indication that upwelling and hence high production is more profound and persistent off Pisco.

Because of the previous correlations in other basins, we can suggest that high biomass is substantial evidence that production off Pisco is greater than elsewhere along Peru. In other words, the benthos is a better indicator of production over a long time period because the highly variable production is integrated over a long time scale, thereby dampening large departure from the average.

Reduced oxygen stress

Gallardo (1963) described three faunal zones off northern Chile. A shallow sublittoral zone conducive to life was found from the shore to a depth of 50 m and seaward of a depth of 400 m was what he described as a typical bathyal fauna. These were separated by a "semiabiotic" region characterized by very low dissolved oxygen in the bottom water and reducing conditions in the sediment. Likewise, off northern Peru, Frankenberg and Menzies (1968) attributed the relatively low biomass at their 519 m station to low bottom water oxygen concentrations. Between these areas, it can therefore be suggested that the stress imposed by lack of oxygen in the bottom water and sediments is the cause of the departure there of the general relationship between depth and biomass. Although the numbers of animals were not markedly affected, biomass in wet weight off Pisco at depths where oxygen values were less than 1 ml/l was depressed by a factor of *ca.* 100 compared to stations near 1000 m. A regression of data from depths greater than 800 m alone follows the pattern predicted from the other areas, but it is elevated far above the other regressions: $Y = 5.5 - 0.00045 (X - \overline{X})(P < 0.01\ b = 0)$. Comparison of this line with a regression from off northern Peru, again utilizing only data from more than 800 mg $[Y = 4.1 - 0.00024\ (X - \overline{X})\ (P < 0.05\ b = 0)]$ suggests, as discussed when considering primary production, that outside stressful low oxygen concentrations, Pisco biomass is far above that to the north. At greater depths this difference appears to be dampened, perhaps as a manifestation of the continued rapid transport of sinking organic material to the north by currents.

CONCLUSIONS

The inverse relationship between the logarithm (base$_{10}$) of biomass and depth can be defined by least squares regressions or estimating equations. Different ocean basins support different amounts of biomass and these differences are reflected in statistically significant differences in the slopes and means of the regressions. The heights of the lines in logarithmic (base$_{10}$) units above the abscissa appear to be related to the magnitude of primary production in a basin, whereas the slopes of the lines appear related to the magnitude of change in primary production from the shallow to the deep, offshore water across the individual regions considered. Depth and primary production in surface water are concluded to be the major factors determining benthic biomass except where the influx of organic material is so high that oxygen depletion causes deleterious stress conditions as off Pisco, Peru. In this situation where dissolved oxygen is the dominant parameter, maximum biomass can be found offshore of the depauperate depths.

Classically, benthic investigations have attempted to estimate food sources for commercially important fish, and although the relationships between benthic biomass and fish populations have never been precisely defined, it must be assumed that areas of highest biomass are characterized by the highest influx of available organic material for food. In this study, it has been demonstrated that this sometimes occurs in areas and depths generally considered beyond the limits of conventional fisheries. In any case, the unexpected pattern outlined relative to low oxygen and surface productivity suggests some intriguing new avenues of exploration and possible exploitation.

Acknowledgements

Support for this investigation came from Contract AT(30-1)-3862 with the U.S. Atomic Energy Commission (Reference NYO-3862-24) and National Science Foundation Grants GZ 259 and GA 1298. Dr. Richard Dugdale is to be thanked for providing time on the *T. G. Thompson*. Drs. John Ryther, David Menzel, Richard Barber, Lou Hobson, and Carl Lorenzen helped interpret the productivity data and supplied recent unpublished data from Peru. Judith I. Rowe aided in sorting the samples and typing the manuscript.

References

BOGDANOV, D. V., SOKOLOV, V. A. and KHROMOV, N. S. (1968). Regions of high biological and commercial productivity in the Gulf of Mexico and Caribbean Sea. *Oceanology*. **8**, 3, 371–381.

CORWIN, N. (1969). Reduced data reports for *Atlantis II* 31, 42 and 48. Tech. Rept. Ref. No. 69-20. Woods Hole Oceanographic Institution. Unpublished manuscript.

FRANKENBERG, D. and MENZIES, R. J. (1968). Some quantitative analyses of deep-sea benthos off Peru. *Deep-Sea Res.* **15**, 623–626.

GALLARDO, A. (1963). Notas sombre la densidad de la fauna bentonica en el sublitoral del norte de Chile. *Gayana Zool.* **8**, 3–15.

MENZEL, D. W. and RYTHER, J. H. (1960). The annual cycle of primary production in the Sargasso Sea off Bermuda. *Deep-Sea Res.* **6**, 351–367.

ROWE, G. T. and MENZEL, D. W. (1969). Quantitative benthic samples from the deep Gulf of Mexico with some comments on the measurement of deep-sea biomass. Manuscript submitted for publication.

RYTHER, J. H. (1963). Geographic variations in productivity. In *The Sea* pp. 347–380. (Ed. M. N. Hill) London: Interscience.

RYTHER, J. H. (1966). Cruise Report, Research Vessel *Anton Bruun*, Cruise 15. Special Rept. Number 5, Marine Laboratory, Texas A and M University.

RYTHER, J. H., MENZEL, D. W. and CORWIN, N. (1967). Influence of the Amazon River outflow on the ecology of the western tropical Atlantic. I. Hydrography and nutrient chemistry. *J. Mar. Res.* **25**, 1, 69–83.

RYTHER, J. H. (1969). Photosynthesis and fish production in the sea. *Science*, **166**, 72–76.

SANDERS, H. and HESSLER, R. (1969). Ecology of the deep-sea benthos. *Science*, **163**, 1419–1424.

SANDERS, H. L., HESSLER, R. R. and HAMPSON, G. R. (1965). An introduction to the study of deep-sea benthic faunal assemblages along the Gay Head-Bermuda transect. *Deep-Sea Res.*, **12**, 845–867.

Enhancement of marine protein production

WALTER R. SCHMITT and JOHN D. ISAACS

Scripps Institution of Oceanography
La Jolla, California 92037

INTRODUCTION

In considerations of artificial upwelling for the enhancement of marine animal protein production for the benefit of humans, it is commonly assumed that a harvest would be increased *in the proportion* of the additional rate of supply of plant nutrients to that originally extant. Isaacs and Schmitt (1969) have postulated possible conditions, however, where nutrient enrichment could lead to a *disproportionately large* stimulation of the food chain, with the net effect of much more additional animal protein available to man. Such extremely favorable conditions might exist in naturally-confined and nutrient-limited waters, as for instance in atoll lagoons, but substantial disproportionate effects may also result from stimulated upwelling in unconfined waters.

As the upwelling agent we will in this study employ the byproduct "heat" from the production of stationary power, turning a potentially harmful effect on the environment into a beneficial one. We will also consider the by-product effects of thermal gradient power (which employs the vast heat energy present in the thermally stratified ocean) as well as the direct use of high-grade power. The potential benefits will be related to man's animal protein needs and stationary power requirements on a per capita basis.

POSSIBLE EFFECTS OF ADDITIONAL NUTRIENTS ON THE FOOD CHAIN

Stable nutrient-limited biological systems are in equilibrium between the input and loss of nutrients in the form of inorganic and organic materials. It is generally pictured that in the marine milieu, between ten and twenty percent of the calorific content of one trophic level appears in the next, while basal and activity metabolism, reproduction, defecation, and retrogressive mortality* account for the large remainder.

An artificial increase in nutrients, as brought about by fertilization or induced upwelling, may affect this equilibrium intermediate to one of three ways:

1. Immediately-increased nutrient losses account for the entire additional input with no significant increase in biological production. This should be the expected effect in those regions and at those times when production is not nutrient-limited.

2. Losses of the increased nutrients occur at nearly the same rate and form as applies to those in the original system. Thus a pro rata population enhancement would become available for man.

3. The additional nutrients are cycled through a food chain in multiple feedback steps free or nearly free of net loss. Thus all or almost all additional input becomes potentially harvestable at any point in the food chain, which, of course, stabilizes at a higher level throughout.

Equilibrium condition (1) is trivial in the context of harvest enhancement and shall concern us no further.

Equilibrium condition (2) is the normal assumption in considerations of the harvest benefits of induced upwelling.

Equilibrium condition (3) may certainly be approached in semiconfined bodies of water where climax predation is carried out by primarily exogenous species. Disproportionately enhanced net production may be passed through the food chain when maintenance nutrient levels are considerably exceeded. This situation is analogous to the productivity increase observable in irrigation agriculture where steep increases are experienced after a given threshold of water supply is exceeded (Schmitt, 1965), particularly when nutrient deficiencies are corrected at the same time.

---

* E.g. mortality leading down in the food web rather than up.

Synergistic effects may also be experienced by unconfined populations. For example, facultative filter-feeding fishes might feed predominantly on phytoplankton after the latter exceeds some lower limit of concentration. O'Connel and Leong at the U.S. Bureau of Commercial Fisheries, La Jolla, in laboratory experiments with *Engraulis mordax* have observed that this fish's dominant feeding mode will change from biting to filtering when the ratio of small to large food organisms exceeds a certain value (personal communication). In this manner a trophic level may be partly bypassed together with its associated high losses. Also, where the natural nutrient supply fluctuates strongly throughout the year, for instance on a seasonal basis, a steadying by supplementation could lead to disproportionate increases in production, especially where the nutrient and insolation regimes are often opposed from the standpoint of biological productivity, as is common in high latitudes.

INDICATOR NUTRIENT

Among the major nutrients essential for life, nitrogen and phosphorus have most often been indicated as growth-limiting (e.g. Sverdrup, *et al.* 1942). The concentrations of both nitrogen and phosphorus compounds generally increase with depth in the ocean. In the artificial upwelling scheme under discussion, both nitrogen and phosphorus would of course be transported into the euphotic zone, along with other naturally occurring nutrients. Because nitrogen fertilizer can be economically fixed from the atmosphere wherever a source of power exists (which is not true of phosphorus despite its wide dispersal in nature), and can be injected into the water along with the upwelling-inducing energy to overcome any differential nitrogen shortage, we will use the concentration of phosphorus in deep water and in living matter to calculate possible animal protein benefits for man.

Each 2 HP of waste heat—this power level will be explained later—will overturn about 3 m³ of water daily if heated by 10°C. This volume contains about 1/4 gram of phosphorus at a concentration of 2.5 μg-atoms/liter, a concentration often encountered at a depth of 500 m.* Using a carbon: phosphorus ratio of 41 : 1 in organic matter and a carbon : dry-weight ratio of 1 : 2.5, the 1/4 gram of phosphorus per day would correspond to

---

* Roels in this Symposium reports that 500 m water gave the highest growth stimulation in algae cultured in waters from the first 1000 m.

25 grams of dry plant material per day on the basis of complete net bio-syntheses. The point of the assumptions of equilibrium condition (3), described above, is that the accumulation of these 25 grams of additional plant material each day would eventually give rise to additional animal material at nearly the same rate in successive trophic levels. Thereby a small amount of *plant* protein will be transformed into a somewhat larger amount of *animal* protein of high biological value by virtue of the general increase in protein concentrations with advancing trophic levels. Allowing some small loss of the additional nutrients, let us say 15 per cent per trophic step in their complex progression through the food chain and hence in their organic equivalents, we would expect the following additions to devolve from the quarter gram of upwelled phosphorus:

Indicator nutrient upwelled: 0.25 gram P/day

Potential net biosyntheses: $0.25 \times 41 \times 2.5 \approx 25$ g OM (dry weight)/day

Primary producers
$$-25 \times 0.85 \approx 21 \text{ g OM/day containing ca. } 6 \text{ g protein}$$

Herbivores $-25 \times 0.85^2 \approx 18$ g OM/day containing ca. 11 g protein

1st Carnicores $-25 \times 0.85^3 \approx 15$ g OM/day containing ca. 11 g protein

2nd Carnivores $-25 \times 0.85^4 \approx 13$ g OM/day containing ca. 10 g protein

HARVESTING AND PROCESSING EFFICIENCIES

In the acquisition of the additional product, we need to differentiate between confined and unconfined upwelling. The former resembles pond culture with the possibility to harvest in full the additional as well as the natural product. Again, some inefficiency in acquiring the product may be unavoidable, and we will set this at 10 per cent. Here, moreover, conditions are favorable for single-species cultivation, thus circumventing the encouragement of unwanted "weed" species at the expense of desirable ones. In the situation of unconfined upwelling we are dealing in essence with a natural open-sea system and are subject to its harvesting limitations. Marine populations have been exposed to a wide range of fishing pressures, from nonuse to near eradication, and it appears that highly reproductive fish species can sustain an average annual harvest rate of about 30 per cent of their biomass (Murphy, 1965), which we accept as a harvesting factor. However, we are only now becoming aware of the consequences when fishing pressure remains insensitive to differences in annual recruitment success which is difficult to quantify, as for instance in the California sardine fishery in the forties and fifties and, more recently, in the North Sea herring fishery. It is possible that supplementation of nutrients by induced upwelling could have stabilized these fisheries and at a higher level of yield. What this high level might be is difficult to say.

We need to account for one more source of inefficiency on the protein's way to humans—the readying of the raw fish for consumption. An as yet experimental preparation, fish flour or Fish Protein Concentrate (FPC), offers almost complete conversion of the raw fish into edible form with a nominal loss of about 15 per cent. Processing seafood in the conventional manner—filleting, canning, drying, etc.—is much more wasteful of animal protein by discarding circa half the fish's weight. These discounts need not apply to the confined case [or to equilibrium condition (3) in general] if the processing wastes are returned to the system.

The foregoing inefficiencies can be combined in numerous ways. We will restrict our consideration to the most favorable combination with the one exception of using two harvesting efficiencies, 90 per cent, and because of the limited opportunities for confined upwelling, 30 per cent in the open sea. Otherwise, the rationale for accepting the most favorable combination is simply that before any large-scale induced-upwelling scheme would be implemented (and its justification depends of course on whether or not the assumptions of disproportionate stimulation of the food chain are in practice realizable, especially in the open sea) the other factors, harvesting and processing, more readily amenable to intervention, will already have been maximized.

PRODUCT AND BY-PRODUCT POWER SOURCES

One of the stimuli toward this study was the realization that the world's stationary power requirements are rising steeply, in fact are rising on a per capita basis. Vast differences exist in its usage among nations. Venezuela, with the highest per capita power production in Latin America, reaches 10 per cent of the United States level. India barely reaches 1 per cent. China might use even less. There is thus plenty of room for expansion if we set an eventual worldwide average level at the United States' present one HP per person.

In the generation of each 1 HP of electrical power by fossil or nuclear

10*

fuels, 2 HP of by-product power are discharged into the environment as heat. Much concern is felt over this waste heat's potentially detrimental impact on the ecology of rivers, lakes, and even of the ocean. We are here suggesting a system that would employ the very much higher level of waste heat of the near future to the benefit of mankind's need for animal protein.

If we look beyond the proven conventional processes for the production of stationary power, we discern the following two strictly marine possibilities. One would exploit the thermal stratification of the ocean, the other the energy of surface waves. The former requires the juxtapositioning of waters of different temperatures under reduced pressure (Romanovsky *et al.*, 1953). On account of the small temperature differential, this process' power-generating efficiency is very low, requiring perhaps 30 times the heat rejection of the conventional processes. Since hydrostatic stability is achieved not only by the heat exchange but also by the addition of about one per cent of fresh water in the condensate, the upwelling by-product is about 60 times greater than that obtainable from a fossil or nuclear-fueled power plant. Not enough is known about the regeneration of the sea's thermal stratification to permit an accurate determination of this energy reservoir's sustainable potential, but it appears to be sufficient to supply 1 HP for perhaps 10^{12} people.

The capture and harnessing of the energy in surface waves has long been an intriguing proposition. One can easily envision a surface-following piston in a fixed pipe supplying power. For the purpose of upwelling one can employ a variant of this system in which the pipe is surface-following while the water within acts as the piston, thus conducting water from depth to the surface. One of us has experimented with this system with encouraging results. If the average seastate can be described as 1 m, 8 sec waves, the energy potential of this source would suffice for 2×10^8 people's stationary power allocation.

Wave power, tides and tsunamis excluded, in essence constitutes a degraded but steadier wind power. Windmills have driven flour mills and water pumps for hundreds of years, and renewed interest is given to this source of power, particularly for marine applications that can tolerate its sporadic regime. Upwelling for fertilization is one such application. While wind power appears to be less abundant than sea thermal power, it could still provide 1 HP for about 10^{10} people, 3/4 which power may be available over the oceans.

POSSIBLE LEVELS OF ANIMAL PROTEIN BENEFITS

We now want to calculate the upwelling benefits, expressed as a proportion of per capita animal needs, if the power sources depicted above were employed to provide an average stationary power supply of 1 HP per person. From various sources (Anon., 1964a and Anon., 1964b) we accept 20 g (dry weight) of animal protein per person and day as a desirable minimum average nutritional level. We have already shown the amount of additional protein that might be stimulated by 2 HP of waste in a highly efficient (85 per cent per trophic step) food chain. For harvesting and processing efficiencies we have accepted 90 and 85 per cent in confined upwelling, and 30 and 85 per cent in unconfined upwelling. We may summarize then the entire sequence of efficiencies as follows:

For confined upwelling—
   $(0.85^n \times \text{protein concentration}) \times 0.90 \times 0.85$,
   for unconfined upwelling—
   $(0.85^n \times \text{protein concentration}) \times 0.30 \times 0.85$,
   where the parenthetic terms remain nearly constant because the protein concentration rises with n.

We obtain thus 8.4 and 2.8 grams, respectively, of consumable animal protein, or 42 and 14 per cent of nutritional need per stationary power allocation (all on a per capita basis) from conventional power sources.

Considering the sea thermal process with 60 times the upwelling by-product, 25 and 8.4 times the nutritional need would be added per power allotment for confined and unconfined upwelling, respectively.

The employment of wind or wave energy for artificial upwelling offers opportunities for spectacular effects, as they represent high-grade product-power free of by-product problems. The power so generated could be used for the mechanical lifting of deep nutrient-laden water to the surface. With the exception of deep-sea regions where heat exchangers could be usefully employed in the intake system (e.g., where salinity decreases with depth), this mechanical system would be limited to pumping into confined bodies of water where there is no hydrostatic necessity to maintain this cold water at the surface by heating. (Warmth may be demanded by the biota, of course, and this would be supplied by insolation and may limit the rates at which cold water may be added.)

The effective lift in this case is composed of the lift from depth D (500 m) against the density differential $\Delta\varrho$ (10^{-3} g/cm^3) and the lift over the enclosure height H (let's say 2 m) at density ϱ (1.025 g/cm³). Some small loss due to friction in the pipe and pump, perhaps 10 per cent, should be allowed for, though its importance would decline as the system becomes larger. With the other assumptions as before, 1 HP of this high-grade power would then stimulate additional animal protein equal to the needs of about 300 people (as FPC in the confined case). Were processing wastes returned to the system, 360 people would be benefited with FPC or conventional seafood. Of course, the high-grade power envisioned here could also come from the product portion of the thermal generators discussed earlier.

SUMMARY

It may be possible that under favorable conditions the marine food chain would utilize artificially-introduced nutrients at a high degree of overall effectiveness. The introduction could be accomplished by the waste heat that appears as by-product in the generation of stationary power, putting a potential environmental hazard to potentially beneficial use. High-grade product power is considered for the special situation of confined upwelling. We have calculated the degree of expectable benefits under some assessment of efficiencies in the total train of events. They may conceivably supply from 14 per cent (unconfined operation) to 36,000 per cent (unconfined operation) of per-capita animal protein needs from the employment of the product and by-product effects of per-capita requirements for stationary power.

References

Anon. (1964a). *Report on Land and Water Development in the Indus Plain*, p 50. The White House, Washington, D. C. 454 pp.
Anon. (1964b). *Recommended Dietary Allowances*, p. 11. NAS-NRC Publ. 1146. Washington, D. C. 36 pp.
ISAACS, JOHN D. and SCHMITT, W. R. (1969). Stimulation of marine productivity with waste heat and mechanical power. *J. Cons. Int. Explor. Mer*, **33**, 1, 20–29.
MURPHY, G. I. (1965). Population Dynamics of the Pacific Sardine, p. 141. Ph. D. Dissertation, Scripps Inst. Oceanog., Univ. Calif., San Diego. 171 pp.
ROMANOVSKY, V., FRANCIS-BOEUF, C. and BOURCART, T. (1953). *La Mer*, p. 116. Larousse, Paris, 503 pp.
SCHMITT, W. R. (1965). The Planetary Food Potential, p. 686. *Ann. N. Y. Acad. Sci.*, **118**, 17, 645–718.
SVERDRUP, H. U., JOHNSON, MARTIN W. and FLEMING, RICHARD H. (1942). *The Oceans*, p. 768. Prentice-Hall, N. Y. 1087 pp.

Fertilization of the sea as a by-product of an industrial utilization of deep water*

ADMIRAL PAULO DE CASTRO MOREIRA DA SILVA

Instituto de Pesquisas da Marinha,
Rio de Janeiro, Brazil

The great efficiency of agriculture, which made civilization possible, was due, among many social and technological factors (property on land, selection and development of highly productive crops, transplantation) to the possibility of enriching soils with fertilizers accumulated by nature and brought to these soils by cheap transportation. In the oceans the nearest possible source of nutrients (if in concentrations infinitely inferior to those encountered in land) and of the still not completely known combination of ingredients (metalic, organic) that favors optimum production is deep water brought to the euphotic layer and there maintained by a shallow shelf. In such a condition repeated re-cycling (through mineralization) would ensure a cumulative effect that could make fertilization self-sustained. Ingredients present in deep water, but sometimes missing in the surface (for instance, silicates), could even favor the substitution of some primary producers by others of higher efficiency.

The aspiration and warming of richer deep waters in the Western Southern Atlantic would not be economically justifiable: such waters are relatively too deep (Fig. 1, Curve A); results—in terms of secondary production—besides being unpredictable, would not be private property, and, so, would not encourage investments. In some favored coastal points, however, cir-

* Short title: Fertilization as a by-product of ice production.

463

culatory factors as well as upwelling may raise and keep the 0.7 µgA/L PO₄—P isoline (corresponding to the 14°C isotherm, to 7 µgA NO₃—N/L and to a fair amount of silicate) at a depth of 50 meters or less (Fig. 1, Curve B). One of these favored positions is Cabo Frio, a bluff 70 miles to the east of Rio de Janeiro (Fig. 2). Downwelling, induced by the south-westerly winds

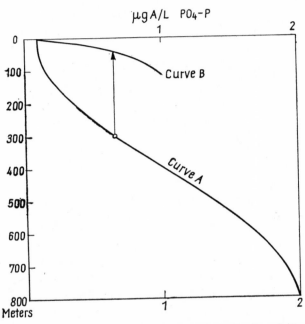

Figure 1 Curve "A" shows the normal vertical distribution of phosphate in the South Atlantic Ocean; curve "B" shows the effect of upwelling near Cabo Frio

associated with the polar invasions, will certainly depress the surface isotherms, but will have practically no effect on the 14°C isotherm. If this 50 meters deep water is aspirated, warmed, and discharged in a secluded bay, it will float and expose to light its nutrients. Under these new conditions a boosted primary production can be expected, if not by some new primary producer, perhaps by diatoms.

The operation given the uncertainty of results, would not be, by itself, economically justifiable: but if the aspiration and warming of this deep water is a free by-product of some profitable industrial utilization of this

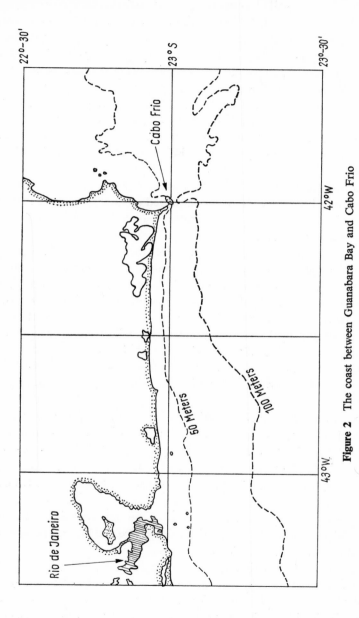

Figure 2 The coast between Guanabara Bay and Cabo Frio

deep-water, then it would constitute an almost free benefit—if any—and an interesting experiment in fertilization. This industrial utilization is the production of ice (worth about 8 US $ per metric ton, in Brazil), concentrated sea-water (up to $70^{o}/_{oo}$ salinity) at 15.6°C, and some fresh water at this same temperature, through the Vacuum-Freezing-Vapor-Compression Process of Desalination developed by Colt Industries in the United States.

7076 metric tons of sea-water (14°C, $35.60^{o}/_{oo}$, 0.7 μgA/L PO_4—P) per day are aspirated from a depth of 40 meters; 1633 tons, after deaeration, are cooled to 2.8°C and then sprinkled into a vacuum chamber maintained at a pressure of 3.4 millimeters. As this is very nearly the triple point, the water will evaporate 46 tons of vapour. The loss of latent heat will cool down the residual water, a part of which will freeze into 266 tons of ice. The remaining 1320 tons will be concentrated to $43.5^{o}/_{oo}$ and, at a temperature of -3.3°C, will cool the aspirated sea-water before it is sprinkled into the chamber (Fig. 3).

To condense the vapour into fresh water, the vapour is first compressed to 4.8 millimeter and then cooled by a conventional ammonia refrigerating unit. Forty six tons of fresh water will be produced, at a temperature of 0.5°C, and will also cool the aspirated sea-water before it is sprinkled into the chamber. Condensation of the ammonia in the unit is ensured by 5443 tons of the aspirated sea-water which will consequently have the temperature raised to 21.7°C, making it sufficiently light to float if discharged into a contiguous bay. As the superficial circulation in this bay has a very slow turn-over, this fertile water will gradually accumulate in the bay.

The daily output of the installation (complete power 600 HP) is 266 tons of ice (worth, in Brazil, some US $ 2,100), 1320 tons of concentrated sea-water ($43.5^{o}/_{oo}$, 15.6°C) useful for cooling purposes and as raw-material for the near-by solar ponds of an alkali industry, 46 tons of fresh-water at 15.6°C, and, of course, 5,443 tons of deep, fertile water at 21.7°C. Different proportions of fresh water, ice and brine (up to a concentration of $70^{o}/_{oo}$) can be produced. The choice of the best combination is purely economical, depending on the relative prices of ice and brine. The by-product is the experimental fertilization of the bay.

The amount of fertilizing water, some 5,500 tons daily, is not very impressive, and it may appear that the same effect could be obtained by simply enriching the bay with conventional fertilizers from land sources. But, besides believing in the higher effectiveness of natural deep-water, its aspira-

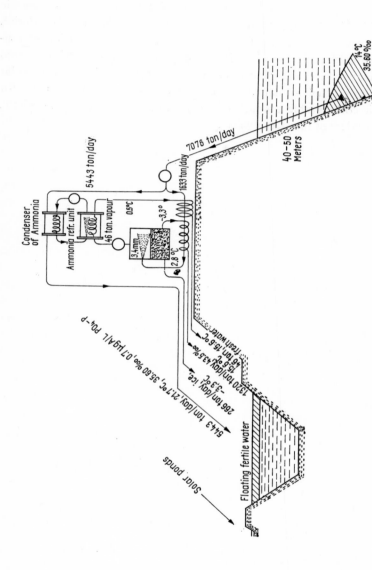

Figure 3 Scheme of the ice, brine, and freshwater factory

tion from 50 meters permits the production of ice with only 41.5 kWh per metric ton, a much better performance than the average 50 kWh per metric ton prevailing in Brazil. And for an industry processing, as raw-material, sea-water naturally evaporated in solar ponds, any concentration of sea-water is a valuable asset, especially if this concentrated sea-water is delivered at a low temperature, since it can also be used, before going into the solar ponds, as an efficient coolant for the thermal systems. In Cabo Frio, for instance, a considerable amount of steam, now lost to the atmosphere, could be condensed into fresh water.

Upwelling and its biological effects in southern Brazil

ADMIRAL PAULO DE CASTRO MOREIRA DA SILVA

Instituto de Pesquisas da Marinha,
Rio de Janeiro, Brazil

Upwelling is so frequently associated in scientific literature with the western coast of the continents that the revelation of the existence of a rather important one in the southern coast of Brazil may come as a surprise. Many years of regular observation and, more recently, the development of a mathematical model by the Instituto des Pesquisas da Marinha demonstrate that not only does this upwelling exists, but that it has a great biological importance in the production and variation of an important fish resource in the country: the sardine (*Sardinella aurita*).

The most interesting fact about this Brazilian upwelling is that as the seasonal cycle of dominant winds is identical with the synoptic weather cycle, and the response of upwelling or downwelling to wind is surprisingly prompt, one has, in some occasions in the span of a week or even less, a contracted view of the succession of events that appear in the seasonal span of one year.

Weather in the southern coast of Brazil is dominated by the presence, east of Rio de Janeiro, of a Tropical Maritime Anticyclone, blowing easterly winds on the coast (Fig. 1). As cold air accumulates in Argentina as a polar anticyclone, the Tropical Maritime Anticyclone retracts, and the winds back to Northeast, North and, finally, Northwest, as the cold front approaches. The passage of the cold front suddenly backs the wind to Southwest, succeeded by South and Southeast as the cold front advances northeastward and the cold anticyclone coversSouth Brazil. Then the cycle repeats itself as the cold polar anticyclone, already in low latitudes, degenerates into a Tropical Maritime Anticyclone.

The cycle is fast in the second half of the year, four days being not an unusual period. In the first half of the year, it is very much slower, since frequently the cold air mass is not strong enough to push the tropical anticyclone into the ocean. The statistical effect is a seasonal one with a dominance of east or northeasterlies during summer and fall and southwesterlies or westerlies during winter and spring.

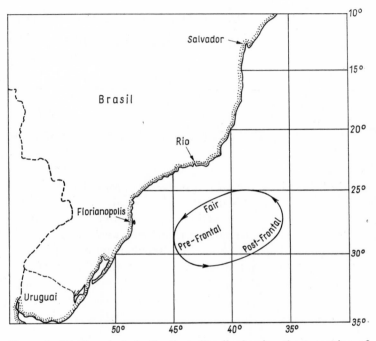

Figure 1 Weather cycle in Southern Brazil, showing the succession of weather types and winds

The synoptic cycle appears, in a cumulative vectorial presentation, as the ellipse seen in Figure 1. The effect of this succession of directions on upwelling and downwelling appears in Figures 2 and 3. In both, the vertical diameter represents the coast and each radius represents the direction from which the wind blows. The curves in the upper figure (Fig. 2) represent the depth (in terms of Ekman's frictional layer, D) of inversion from convergence to divergence. The curves in Figure 3 represent the amount of

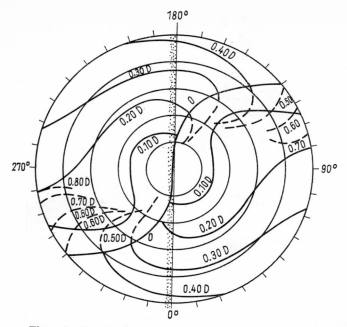

Figure 2 Depth of thermocline for winds of different incidences

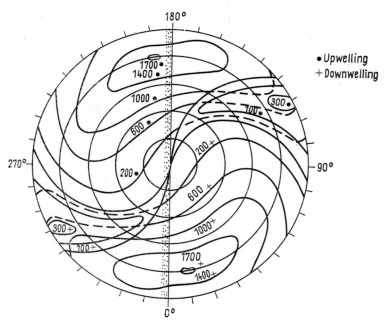

Figure 3 Vertical fluxes for winds of different incidences

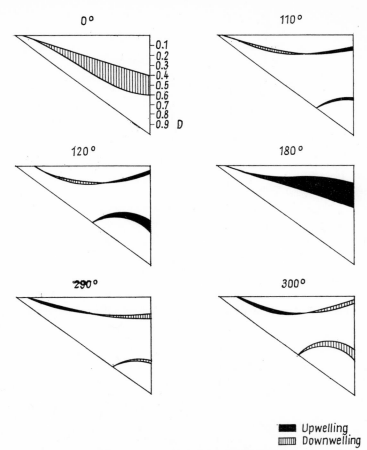

Figure 4 Profiles showing upwelling and downwelling for winds of different incidences

upwelling or downwelling in relative units. Figure 4 shows upwelling and downwelling in successive phases of the wind gyre.

As the effect of a complete gyre (either in the synoptic short scale, or in the statistical yearly cycle) can be computed for different stretches of the coast, it is apparent (Fig. 5) that in all stretches (considering the usual velocities of the different winds) about three days of fair weather are necessary to dissipate the downwelling effect of a frontal invasion. Also, the two maxima in the production of sardine occur in coincidence with the two maxima of seasonal upwelling (Fig. 6).

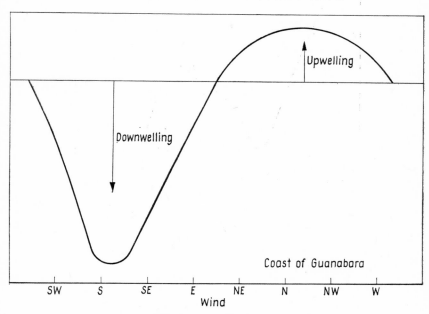

Figure 5 Relative upwelling and downwelling for winds from different directions

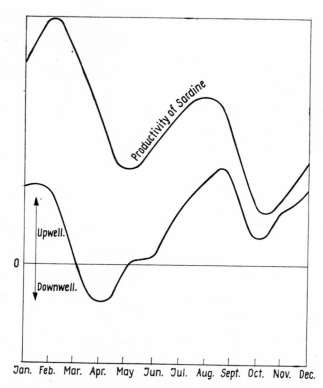

Figure 6 Correlation of upwelling and production of sardine in Rio area

Chlorophyll a, carbon, and nitrogen in particles from a unique coastal environment

LAWRENCE F. SMALL

and

DONALD A. RAMBERG

Department of Oceanography
Oregon State University
Corvallis, Oregon USA, 97331

Abstract

Off Oregon U.S.A. in summer a strong seasonal pycnocline near the sea surface (from Columbia River discharge and surface heating) tends to intercept the seaward movement of water upwelled along the coast from below a permanent pycnocline at about 75 m depth. The consequence of this situation is often the sinking of the surfaced upwelled water and its subsequent movement offshore between the two pycnoclines. Other workers have demonstrated this circulation pattern with hydrographic and turbidity measurements; we have demonstrated that biological quantities such as chlorophyll a, particulate nitrogen (N), a particulate carbon-nitrogen ratio (C/N), and apparent oxygen utilization (AOU) behave in a similar manner. The scheme is often encountered, but apparently is not an obligate one off the coast, as measured by biological quantities. Weakened gradients between surface upwelled water and water above the seasonal pycnocline sometimes will allow the upwelled water to continue offshore at the surface, with the result that relatively large phytoplankton populations develop there.

Upwelling is not a continuous process along the coast in summer, and it appears that populations of phytoplankton tend to build up between actual loci of upwelling; that is, in longshore profile chlorophyll *a*, N, C/N, and AOU indicate smaller standing stocks directly over upwelling domes and larger stocks between the domes. Horizontal surfaces at three different depths over the whole coastal area reveal the general paucity of viable phytoplankton in surface waters above the seasonal pycnocline (except within about 75 km of the Columbia River mouth) and in waters below the photic depth (about 20 m).

INTRODUCTION

The summer coastal waters off Oregon, U.S.A. are ecologically unique in that two rather distinct hydrographic processes occur there (Pattullo and Denner, 1965). As a consequence of prevailing winds from the north and northwest, relatively cold, saline, nutrient-rich water is upwelled along the coast from below a permanent pycnocline (60–100 m depth), and relatively warm, fresh, nutrient-impoverished water extends southward along the coast in a plume from the Columbia River. Solar heating of the river discharge as it moves away from the river mouth and mixes with seawater serves to maintain a strong seasonal pycnocline at 15–30 m.

From hydrographic and current velocity profiles in the Oregon coastal region, Mooers and Smith (1967) proposed that water upwelled near the coast from below the permanent pycnocline begins to flow offshore along the surface until its relatively high density causes it to sink beneath the seasonal pycnocline. The sinking water then continues to flow offshore along the permanent pycnocline. This model is supported by Pak, Beardsley, and Smith (in press), who followed a temperature inversion and turbidity maximum through the critical region of apparent sinking. Direct current measurements (Collins, *et al.* 1968) have indicated that the mean flow above the permanent pycnocline during upwelling is mainly in a southerly, longshore direction, however, although there is an offshore component.

In this paper we test the relevance of the proposed circulation model to certain biological and chemical distributions. Vertical profiles of chlorophyll *a*, particulate nitrogen, apparent oxygen utilization (AOU), and a particulate carbon-nitrogen ratio are used. In addition, possible effects of longshore water movements on the above biological quantities are studied, and horizontal distributions of the quantities are examined over the total sampling area at three depths in order to biologically characterize plume waters and upwelled waters.

MATERIALS AND METHODS

Eighty-seven stations were occupied off the Oregon coast between 24 June and 3 July, 1968. The dates of sampling were selected to correspond with a time of strong development of the Columbia River plume off Oregon (Anderson, *et al.* 1962) and with incidence of well-developed coastal up-welling. At each station water samples for chlorophyll *a* and particulate carbon and nitrogen analyses were retrieved with non-toxic collecting bottles from the surface and from 10 and 50 m depths. Twenty samples to 200 m (depth permitting) were taken at each station for dissolved oxygen, nitrate, and salinity. Dissolved oxygen was determined by Winkler titration, nitrate by Autoanalyzer®, and salinity by inductive salinometer, all on board ship. Apparent oxygen utilization (AOU) was computed as the difference between the original oxygen concentration at the time the water sample was withdrawn. Redfield, Ketchum, and Richards (1963) discuss the assumptions and errors involved in AOU determination and suggest that the errors are not sufficiently large to invalidate the use of AOU as a tool to examine biological activity in the sea. Changes in AOU in surface waters require careful interpretation, however, as oxygen exchange with the atmosphere occurs in addition to photosynthetic production of oxygen.

Samples for chlorophyll *a* determination were membrane-filtered and stored in a desiccator at −20°C on board ship, and then analyzed at the shore-based laboratory using procedures given by Strickland and Parsons (1968). Samples for particulate C and N determinations were filtered through borosilicate fiber filters and placed in a desiccator at −20°C on board ship. Analyses were made in the shore-based laboratory using an F and M Carbon–Hydrogen–Nitrogen Analyzer standardized with acet-anilide of known C–H–N composition.

RESULTS

Vertical profiles are shown for the Depoe Bay (DB) sampling line on 24 and 28 June (Figs. 1 and 2). Density (σ_t) has been plotted (Fig. 1a, d) in order to show the spatial relationship between the permanent and seasonal pycno-cline, and nitrate (NO_3) is plotted (Fig. 1b, e) to indicate the relationship of this limiting nutrient to upwelled water. Collins (1964) has defined the permanent pycnocline off Oregon as the 25.5–26.0 σ_t band. As the pycno-cline is inclined toward the sea surface near shore during upwelling, the

25.5 σ_t contour can be used conservatively to designate the leading edge of water upwelled from below about 75 m. The 25.5 σ_t line moves toward the surface most strongly at stations DB 5 and DB 7, between 24 and 28 June (Fig. 1a, d). The maximum rate of vertical displacement at DB 5 is approximately 3×10^{-3} cm/sec, which is comparable to the rates found off the extreme southern Oregon coast by Smith, Pattullo, and Lane (1966).

We have adopted the 23.5 σ_t contour to delineate the seasonal pycnocline (Fig. 1a, d). This contour at the surface corresponds to the $32.0^o/_{oo}$ salinity contour and to the 0.127 specific alkalinity contour (Park, 1966), and thus is a more conservative measure of Columbia River plume waters than the $32.5^o/_{oo}$ surface salinity boundary usually given (Anderson *et al.* 1962).

Features similar to those in σ_t and NO_3 sections are observed in the AOU profiles (Fig. 1c, f). Also conspicuous in each AOU transect is a core of negative values which lies mainly below the 23.5 σ_t contour offshore and tends to surface at stations DB 7 to DB 10. On 24 June, the highest concentrations of chlorophyll *a* and particulate N, and the lowest C/N ratios, are distributed rather uniformly inshore of DB 7, although a surface band of relatively high N and low C/N exists between DB 15 and DB 30 (Fig. 2a, b, c). On 28 June, however, the major features are all inshore of DB 10, and the contours have changed markedly (Fig. 2d, e, f).

Figures 3 and 4 show distributions of the quantities in a northsouth direction along the coast. The stations chosen to make up the longitudinal transect are all those which lie within about 3.5 km of shore. Although the transect is a composite picture over four days of sampling, and therefore has little fine detail, the large persisting features are evident. Four general regions of doming of isolines are marked by arrows, and are particularly evident in the σ_t, NO_3 and AOU sections (Fig. 3). The 25.5 σ_t contour breaks the surface only in the southern portion of the transect, and the 23.5 σ_t line appears only in the northern portion at the surface (Fig. 3a). It should be noted that at DB 1 the 26.0 σ_t contour does not follow the 25.5 σ_t line, so that the interpretation of upwelling at DB 1 might be in error. A region of very low nitrate and large negative AOU values is observed at the surface between stations MC 50 and MC 52, and a similar though less striking region is seen at the surface between MC 25 and MC 29 (Fig. 3b, c). Generally, high N and low C/N values are associated with these pockets of low NO_3 and negative AOU, but chlorophyll *a* distribution bears no consistent relationship (Fig. 4a, b, c). Concentrations of chlorophyll *a* above 0.50 mg/m³ are found only in the southern half of the transect, near the surface.

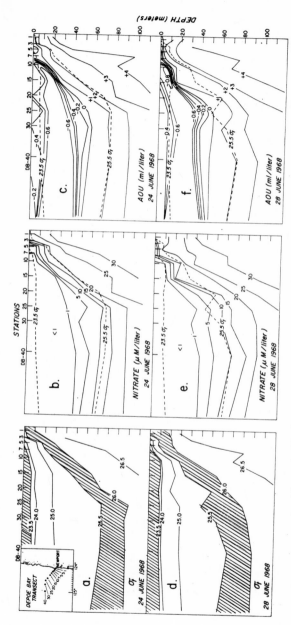

Figure 1   Vertical sections of σ_t, nitrate, and AOU off Depoe Bay, Oregon on two dates. The 23.5 σ_t line represents the bottom edge of the seasonal pycnocline, and the 25.5 σ_t line represents the upper edge of the permanent pycnocline

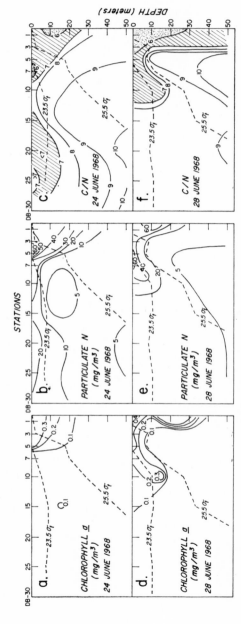

Figure 2 Vertical sections of chlorophyll *a*, particulate N, and C/N off Depoe Bay, Oregon on two dates. The 23.5 σ_t line represents the bottom edge of the seasonal pycnocline, and the 25.5 σ_t line represents the upper edge of the permanent pycnocline

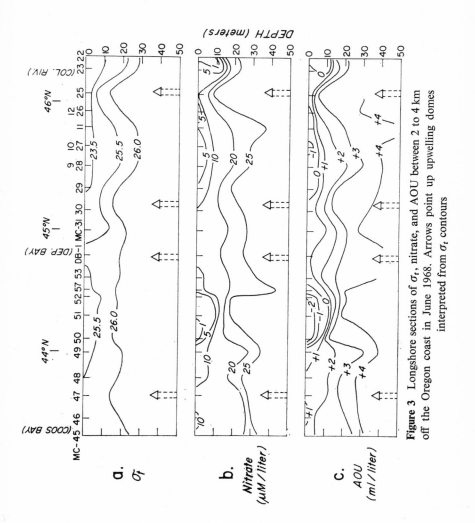

Figure 3 Longshore sections of σ_t, nitrate, and AOU between 2 to 4 km off the Oregon coast in June 1968. Arrows point up upwelling domes interpreted from σ_t contours

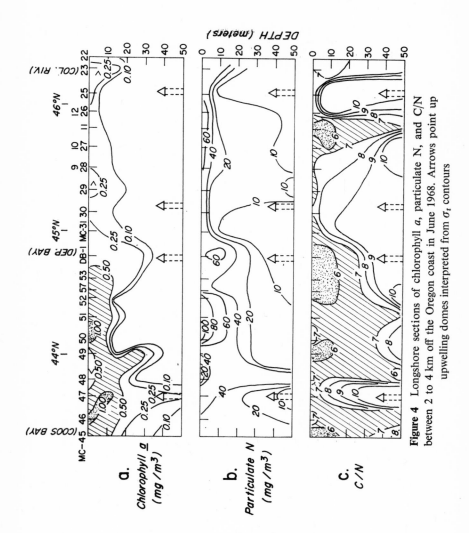

Figure 4 Longshore sections of chlorophyll *a*, particulate N, and C/N between 2 to 4 km off the Oregon coast in June 1968. Arrows point up upwelling domes interpreted from σ_t contours

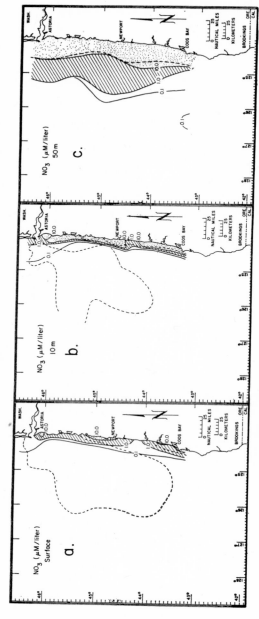

Figure 5 Horizontal sections showing the distributions of nitrate and Apparent Oxygen Utilization (AOU) at the sea surface, 10 m, and 50 m off the Oregon coast in June 1968. The dashed (---) line contiguous with the Columbia River mouth represents the 23.5 σ_t contour, while the dashed contour(s) adjacent to the coast represent upwelled water (25.5 σ_t)

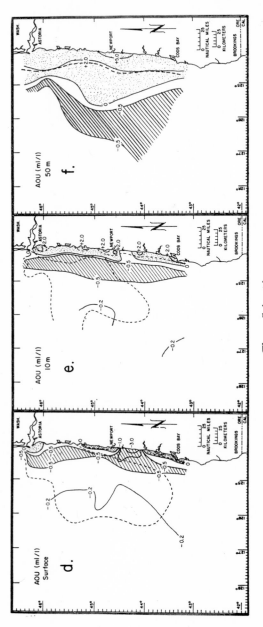

Figure 5 (cont.)

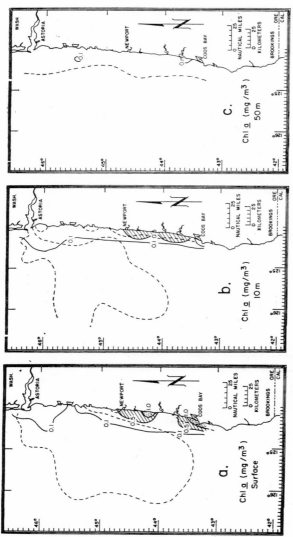

Figure 6 Horizontal sections showing the distributions of chlorophyll *a*, particulate nitrogen, and the particulate carbon-nitrogen ratio (C/N) at the sea surface, 10 m, and 50 m off the Oregon coast in June 1968. The dashed (--) line contiguous with the Columbia River mouth represents the 23.5 σ_t contour, while the dashed contour(s) adjacent to the coast represent up-welled water (25.5 σ_t).

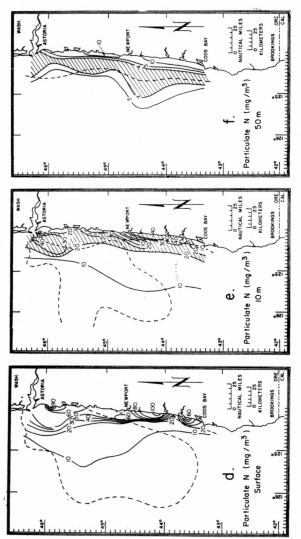

Figure 6 (cont.)

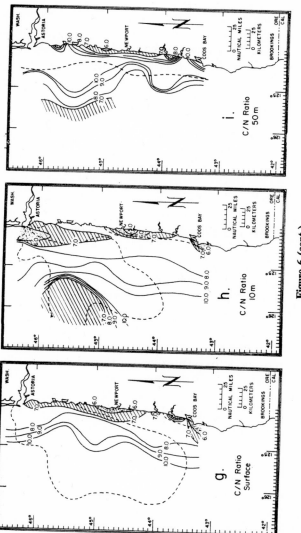

Figure 6 (cont.)

Horizontal surfaces at the sea surface and at 10 and 50 m (Figs. 5 and 6) show the distributions of five quantities in relation to the 23.5 and 25.5 σ_t contours. It is evident in the NO_3 and AOU surfaces that the rather widely separated contours at 50 m become restricted to a narrow coastal band at 10 m and the sea surface (Fig. 5). The core of negative AOU is more easily visualized as a trough in these figures. Chlorophyll *a* is measureable only along the coast and mainly outside the 23.5 σ_t contour, with highest concentrations at or near the sea surface in the southern part of the sampling area (Fig. 6a, b, c). Areas of high concentration of N and low C/N values generally, but not exclusively, correspond to areas of high chlorophyll *a* concentration in the upper 10 m (Fig. 6). An apparent ridge of high N at 10 m (Fig. 6e) appears to lie over a trough of low N at 50 M (Fig. 6f).

DISCUSSION

Evidence in support of the circulation pattern suggested by Mooers and Smith (1967) comes largely from interpretations of vertical profiles of AOU, chlorophyll *a*, particulate N, and C/N ratios on 28 June (Figs. 1 and 2). It is evident that as water is upwelled along the coast, AOU values proceed from positive to negative (mainly due to atmospheric oxygenation), and chlorophyll *a* and N concentrations increase as the nutrient-rich water moves up into the photic zone. The photic depth (depth of 1% surface radiation) is about 20 m in the area DB3 to DB 7. As the water moves offshore at the surface, increasing photosynthetic production of oxygen reinforces atmospheric oxygen exchange to maintain the slightly negative AOU values. The developing phytoplankton populations are then apparently entrained below the seasonal pycnocline, as indicated by the shapes of the chlorophyll *a*, N, and C/N profiles on 28 June (Fig. 2d, e, f). A negative AOU core is developed between the seasonal and permanent pycnoclines largely as a result of continued photosynthetic oxygen production and reduced oxygen exchange through the pycnoclines (Fig. 1f). A characteristic value in the AOU core is -0.6 (although a bolus of -0.8 water is seen in the core near the surface), while in surface water a characteristic value is -0.2.

The chlorophyll *a* and particulate N maxima, and the C/N minimum, extend only a short distance into the AOU core on 28 June, which suggests that maintenance of the AOU core seaward of about 20 km is by *in situ* production from smaller phytoplankton stocks. Pak, Beardsley, and Smith (in press) also noted that their turbidity maximum, which they suggested was the result biological production, extended only a limited distance (about 13 km) from the coast. Anderson (1969) has demonstrated the existence of a photosynthetically viable subsurface chlorophyll maximum in waters off the Oregon coast, and although this maximum is located about 60–75 m deep at 225 km offshore, it is feasible that our AOU core (seaward of about 20 km) is an inshore extension of this photosynthetically active zone. Anderson suggests that the subsurface maximum, deduced in part from oxygen data as well as chlorophyll, may extend over large areas of ocean. Our relatively broad negative AOU trough seen in horizontal section at 50 m (Fig. 5f) is probably a slice through the subsurface maximum.

Distributions of all quantities on 24 June off Depoe Bay (Figs. 1 and 2) suggest that sinking below the seasonal pycnocline is not necessarily a persistent event. It should be noted that the negative AOU core is weakened near the surface relative to that on 28 June, and that there are no extensions of chlorophyll *a*, N, or C/N into the core (Fig. 1c). In fact, the band of relatively high N and C/N at the surface between DB 15 and DB 30 probably is a remnant of biologically productive inshore water that has been moved offshore. Upwelled water moving directly offshore above the seasonal pycnocline might also maintain substantial photosynthetic oxygen production, but would be subject to continuous exchange with the atmosphere as well. The net effect would be the maintenance of near-zero AOU values at the surface. That oxygen exchange is an extremely significant process in Oregon coastal waters has been amply demonstrated by Pytkowicz (1964). He found that the total oxygen exchange with the atmosphere, from April to September, almost doubled the total photosynthetic input in the top 45 m during the same period of time. Water moving directly offshore between the seasonal and permanent pycnoclines would develop a negative AOU core as the result of *in situ* primary production with little oxygen exchange through the pycnoclines.

Although more data are required to verify the hypothesis, it appears that large populations of phytoplankton tend to develop, or be moved, between loci of strongest upwelling (Figs. 3 and 4). For example, the apparently productive surface pockets between MC 50 and MC 52 and between MC 25 and MC 29 are located between σ_t and NO_3 domes. The pocket between MC 50 and MC 52 is not well defined in the C/N data (Fig. 4c), and the pocket between MC 25 and MC 29 is absent in the chlorophyll *a* data (Fig. 4a); however, consistency in the NO_3, AOU, and N contours seems to point to significant biological activity in both areas. At present, two explanations seem possible for the occurrence of large populations between areas of active upwelling. First, it is conceivable that the offshore movement of upwelled water from areas where no large populations develop is fast enough to preclude significant cell division adjacent to the coast. Mooers *et al.* (1968) have shown that current speeds of about 20 cm/sec can persist for several hours in the top 10 m under upwelling conditions, which is clearly fast enough to prevent development of a substantial population. Speeds of 1 or 2 cm/sec., however, even if they persist for a day or more in an offshore direction, will allow a population to develop. The populations which develop between the upwelling loci, then, are possibly nurtured from slow input of nutrients from weak upwelling areas or from land into relatively motionless or slowly circulating pockets between the major upwelling areas.

A second explanation for the apparent development of phytoplankton stock between upwelling loci is the possible diffusion of slowly upwelled water in a longshore direction, to create nutrient-rich pockets between the zones of offshore movement. Some suggestion that slow, longshore circulation might occur in these pockets is seen in the longshore chlorophyll *a* contours (Fig. 4a). The tongue of relatively high chlorophyll *a* concentration extending from about 15 to 40 m depth at station MC 49 apparently is a real feature, as it corresponds to downward extensions in the σ_t, NO_3, and AOU contours at MC 49 (Fig. 3). A circulation pattern possibly is established in which water upwells at stations MC 46 and MC 47 and sinks at MC 48 and MC 49. If sinking does occur, some of the pigments from the highly productive region MC 50 to MC 52 would likely be entrained into the region at MC 49, also. These same features do not appear as clearly in the particulate N and C/N data.

From horizontal sections it is evident that biological quantities in upwelled water and in water above the seasonal pycnocline are different, except around the mouth of the Columbia River (Figs. 5 and 6). Data at 50 m are characteristic of upwelled water below the seasonal pynocline and out of the photic zone; that is, no chlorophyll *a* is evident (Fig. 6c) but a trough of low N (<4.0 mg/m^3) and a ridge of high C/N (<10.0) develop along the 25.5 σ_t contour (Fig. 6f, i). Such values are associated with high nitrate concentrations (Fig. 5c) and positive AOU values (Fig. 5f). As the water moves into the photic zone, biological production turns the trough of low N into an apparent ridge of high N (<20.0 mg/m^3) at 10 m (Fig. 6e). Columbia River discharge begins to affect distributions along the northern part of the coast, as exemplified by the slight bulges in the chlorophyll *a* (Fig. 6b) and N contours at 10 m, and the wedge of low C/N (<7.0) continguous with the river mouth (Fig. 6h). Nitrate is reduced to less than 0.1 mg/m³ (Fig. 5b) and a broad negative AOU trough is maintained a slight distance from the coast (Fig. 5e).

At the surface the strong bulging of contours for particulate N off the river mouth (Fig. 6d), and to a lesser degree for chlorophyll *a* (Fig. 6a) and C/N (Fig. 6g), attest to the river effect along the northern coast. The high particulate N and low C/N values, coupled with relatively low chlorophyll *a* values, suggest that the particles discharged with the river water are largely non-chlorophyllous biological fragments, probably of terrestrial origin. In support of this suggestion, Small and Curl (1968) demonstrated that light extinction per unit of chlorophyll *a* is significantly greater within about 75 km of the Columbia River mouth in summer than anywhere else along the Oregon coast. In plume water south of the immediate influence of river discharge chlorophyll *a* is rarely found in concentrations greater than 0.1 mg/m³, particulate N is less than 20 mg/m³, and C/N is greater than 8.0. Our data show an apparent ridge of high C/N (>10.0) at 10 m that is more or less oriented along the plume axis (Fig. 6h).

CONCLUSIONS

We have demonstrated that biological quantities such as chlorophyll *a*, particulate N, C/N, and AOU can behave in a manner consistent with the circulation scheme proposed by Mooers and Smith (1967), but need not always behave in this manner. Weakened gradients between surface upwelled water and water above the seasonal pycnocline sometimes will allow the up-welled water to continue offshore at the surface rather than sink between the seasonal and permanent pycnoclines. Also, because upwelling is continuous neither in space nor time along the coast in summer, phytoplankton populations apparently build up between active upwelling loci; that is, in longshore profile chlorophyll *a*, N, C/N, and AOU indicate smaller standing stocks directly over upwelling domes and larger stocks between the domes. The general paucity of viable phytoplankton in surface waters above the seasonal pycnocline (except near the Columbia River mouth) and in waters below the photic zone is also observed.

Both horizontal and longshore transects indicate in a general way the relationships of biological properties in upwelled and plume waters off

12*

Oregon. However, because of time lags in sampling and in development of biological populations, fine scale relationships are masked. Vertical transects over shorter time intervals show more detail, but are not close enough together to predict rates of change. Current research is being directed toward time series data collection to analyze the relationships between rates of change of quantities.

This research was supported by PHS grant ES00026. We wish to acknowledge the support given by ship's personnel during the time at sea.

References

ANDERSON, G. C. (1969). Subsurface chlorophyll maximum in the northeast Pacific Ocean. *Limnol. Oceanog.* **14**, 386–391.

ANDERSON, G. C., BANSE, K., BARNES, C. A., COACHMAN, L. K., CREAGER, J. S., GROSS, J. M. and MCMANUS, D. A. (1962). Columbia River effects in the Northeast Pacific. *Dept. Oceanography, Univ. Washington, Seattle* (*Ref. M62-5, Mimeographed*), 17 pp.

COLLINS, C. A. (1964). Structure and kinematics of the permanent oceanic front off the Oregon coast. *M. S. Thesis, Oregon State Univ. Corvallis.* 53 pp.

COLLINS, C. A., MOOERS, C. N. K., STEVENSON, M. R., SMITH, R. L. and PATTULLO, J. G. (1968). Direct current measurements in the frontal zone in a coastal upwelling region. *J. Oceanog. Soc. Japan*, **24**, 295–306.

MOOERS, C. N. K. and SMITH, R. L. (1967). Dynamic structure in an upwelling frontal zone (abstract), *Trans Am. Geophys. Union* **48**, 125–126.

MOOERS, C. N. K., BOGERT, L. M., SMITH, R. L., and PATTULLO, J. G. (1968). A compilation of observations from moored current meters and thermographs (and of complementary oceanographic and atmospheric data. *Dept. Oceanography, Oregon State Univ., Corvallis, Data Rpt.* **30** (*Ref. 68-5, Mimeographed*), 98 pp.

PAK, H., BEARDSLEY, G. F., JR. and SMITH, R. L. (in press). An optical and hydrographic study of a temperature inversion off Oregon during upwelling. *J. Geophys. Res.*

PARK, K. (1966). Columbia River plume identification by specific alkalinity. *Limnol. Oceanog.* **11**, 118–120.

PATTULLO, J. and DENNER, W. (1965). Processes affecting seawater characteristics along the Oregon coast. *Limnol. Oceanog.* **10**, 443–450.

PYTKOWICZ, R. M. (1964). Oxygen exchange rates off the Oregon coast. *Deep-Sea Res.* **11**, 381–389.

REDFIELD, A. C., KETCHUM, B. H. and RICHARDS, F. A. (1963). The influence of organisms on the composition of seawater. In *The Sea, v.* 2, 26–77. (Ed. Hill, M. N.) New York: Interscience Press.

SMALL, L. F. and CURL, H. C., JR. (1968). The relative contribution of particulate chlorophyll and river tripton to the extinction of light off the coast of Oregon. *Limnol. Oceanog.* **13**, 84–91.

SMITH, R. L., PATTULLO, J. G., and LANE, R. K. (1966). An investigation of the early stage of upwelling along the Oregon coast. *J. Geophys. Res.* **71**, 1135–1140.

STRICKLAND, J. D. H. and PARSONS, T. R. (1968). *A Practical Handbook of Seawater Analysis.* Ottawa: The Queen's Printer.

Further enrichment experiments using the marine centric diatom Cyclotella nana (clone 13-1) as an assay organism

THEODORE J. SMAYDA

Graduate School of Oceanography
University of Rhode Island
Kingston, Rhode Island 02881

Resumen

É discutido o crescimento da Cyclotella nana (clone 13-1), incubada a 600 candelas/pé e 20°C durante cinco dias, em 168 experimentos de enriquecimento usando água filtrada coletada da superfície, 50 e 100 metros em 40 estações oceânicas entre Bermuda e Porto Rico, em novembro de 1963. É demonstrada a presença de água de "má qualidade" a 100 metros em todas as estações resultando na mortalidade do inoculo, quando não enriquecido, como tembém a influência de vários enriquecimentos com nutriente (14 rações) eliminando êsse efeito deletério, sua influência no crescimento de *Cyclotella* e suas interações.

São discutidos alguns problemas ecológicos gerais sugeridos pelos experimentos de ensaios biológicos de enriquecimento.

INTRODUCTION

The growth of the assay diatom *Cyclotella nana* (clone 13-1) in nutrient enriched water collected during August, 1963 at three depths from the upper 100 m at four stations (Fig. 1) between Narragansett Bay, Rhode Island and Bermuda has been reported (Smayda, 1964). It was found that 1. nutrients (13 rations) added to surface waters invariably inhibited growth relative to the unenriched control; 2. nutrients added to deeper waters, especially 100 m, frequently promoted growth; and 3. silicate enrichment promoted growth more frequently and to a greater degree than other nutrients when added singly.

This remarkably consistent response over an area approximately seven degrees of latitude contrasts with the more pronounced regional variations in the pattern of sea water quality detected in the North Sea and North Atlantic (Johnston 1963, 1964), and in Puerto Rican coastal waters (Smayda, 1970). *Cyclotella nana* (13-1) assay experiments carried out during November 1963 (Cruise "Colombo") between Narragansett Bay and Puerto Rico to confirm and extend the August bioassay results will be reported here.

MATERIALS AND METHODS

The station locations and recorded hydrographic conditions are presented in Figure 1 and Table 1. Because of foul weather only station 4 of the four August (Cruise "Leo") stations could be sampled; stations 5, 7 and 9 between Bermuda and Puerto Rico were not sampled during that cruise. The thermocline was at 35 m at station 9, and at 65 m at the other stations. The temperature progressively increased southwards toward Puerto Rico; somewhat higher nutrient concentrations were found at station 7.

The bioassay procedure during *Colombo* was slightly modified from the August experiments (Smayda, 1964). Water samples collected from the surface, 50 and 100 m were filtered through a Gelman Type A glass fiber filter *to remove the natural plankton and detrital population*; 15 ml aliquots of the *filtered* water were then dispensed aseptically into transparent 30 ml pre-sterilized Falcon disposable, styrene tissue culture flasks. At each depth 14 different aseptic nutrient enrichments were then made as follows:

1. Enriched with a Basic Inorganic Medium (BSM) containing:
   a) $NaH_2PO_4 \cdot H_2O$ 5 µgA P/L
   b) $NaNO_3$ 50 µgA N/L
   c) $Na_2SiO_3 \cdot 9 H_2O$ 50 µgA Si/L
   d) Fe-EDTA (ferric sequestrene)   50 µg Fe/L
2. $BSM-P-PO_4$ (i.e. N + Si + FeEDTA)
3. $BSM-NO_3-N$ (i.e. P + Si + FeEDTA)
4. $BSM-Si-SiO_3$ (i.e. P + N + FeEDTA)
5. BSM-FeEDTA (i.e. P + N + Si)

6. BSM plus a Trace Metal (TM) mix containing:
  a) $CuSO_4 \cdot 5 H_2O$ 5 μg Cu/L
  b) $ZnSO_4 \cdot 7 H_2O$ 10 μg Zn/L
  c) $CoCl_2 \cdot 6 H_2O$ 5 μg Co/L
  d) $MnCl_2 \cdot 4 H_2O$ 100 μg Mn/L
  e) $Na_2MoO_4 \cdot 2 H_2O$ 5 μg Mo/L
7. BSM plus a Vitamin mix (VITS) containing:
  a) B_{12} 1.0 μg/L   b) Biotin   1.0 μg/L
  c) Thiamine · HCl   0.2 mg/L

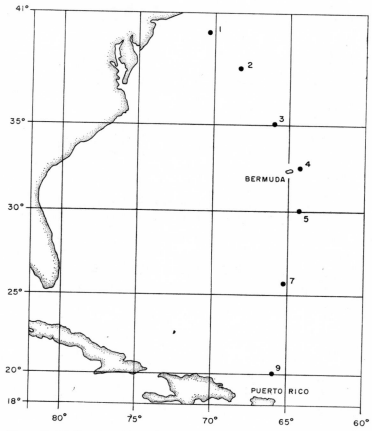

Figure 1 Station locations where *Cyclotella nana* (clone 13-1) bioassay studies were conducted. Stations 4–9 were occupied during R/V TRIDENT Cruise *Colombo* and stations 1–4 during Cruises *Leo* and *Nest*

Table 1 Position of the stations and hydrographic conditions where enrichment experiments were conducted

Station	Position		Depth (m)	°C	⁰/oo	O_2 (ml L⁻¹)	PO_4–P (µg A L⁻¹)	NO_3–N (µg A L⁻¹)	SiO_2 (µg A L⁻¹)
4	32° 22′ N	64° 15′ W	0	23.23	36.425	4.85	0.045	*	0.15
			50	23.25	36.499	4.79	0.010	*	0.30
			100	19.11	36.572	4.55	0.102	*	0.90
5	30° 00′ N	64° 35′ W	0	24.44	36.624	4.75	0.102	0.24	0.30
			50	24.46	36.691	4.71	0.045	0.14	0.15
			100	20.37	36.679	4.94	0.025	0.03	0.30
7	25° 37′ N	65° 12′ W	0	26.00	36.614	4.53	0.160	0.63	0.75
			50	25.90	36.699	4.51	0.060	*	0.75
			100	21.55	36.701	4.86	0.060	*	0.94
9	19° 57′ N	65° 55′ W	0	28.94	36.147	4.34	0.030	*	—
			50	27.19	36.639	4.85	0.010	0.21	—
			100	24.09	36.782	4.86	0.066	0.07	—

(—) no observations made; (*) amount below that detectable by analytical method used.

8. BSM + TM + Vitamins (ALL)
9. ALL—P
10. ALL—N
11. ALL—Si
12. ALL—FeEDTA
13. ALL—FeEDTA + Na$_2$EDTA (1 mg/L)
14. No nutrients added (NONE)

Stock cultures of axenic *Cyclotella nana* (13-1), originally isolated from the Sargasso Sea by Guillard (Guillard and Ryther, 1962) were maintained on board ship in enriched Narragansett Bay water (ration No. 8). For use in the biossay experiments, log-phase *Cyclotella* cells were first transferred from Narragansett Bay water into sterile, filtered, unenriched Sargasso Sea water to deplete their nutrient reserves and induce nutrient starvation and, secondarily, to minimize nutrient carry over with the inoculum. (A second transfer into unenriched Sargasso Sea water did not produce a sufficient number of cells for inoculation into the assay flasks.) The inoculum (based on a haemocytometer count), usually taken from a 7 day old culture, varied from 5480 to 7040 cells per ml; it was constant for all depths at a given station. The culture flasks were incubated for five days at 20 C and about 600 foot candles of continuous artificial illumination. The morphological and physiological-ecological characteristics of *Cyclotella nana* (13-1) have been summarized elsewhere (Smayda, 1964). The experiments were terminated by fixing with a few drops of Lugol's solution, and the population enumerated ashore by making replicate counts with a haemocytometer. Lugol's preservative is inferior to neutralized formalin when added to the Falcon flasks. The strength of the preservative considerably attenuates with time, partly through absorption of the iodine by the plastic. The flasks also leaked on several occasions when dissolution of the sealed edges occurred, possibly caused by the glacial acetic acid used in the preservative. The change in cell numbers relative to the inoculum size, or to that with no (NONE), or to complete enrichment (ALL) will be used as a measure of growth, as specified in the text.

RESULTS

Growth in unenriched water

Significant vertical and regional differences occurred in the natural potential of *unenriched* water to support growth of *Cyclotella nana* (13-1) (Table 2).

Table 2 Growth of *Cyclotella nana* (clone 13-1) in unenriched samples (C) relative to inoculum concentration (I) expressed as C-I (cells/ml) and C/I, and accompanying phosphate concentrations (μg A L^{-1}) during cruises *Colombo, Leo and Nest*

COLOMBO (4–8 November 1963)

Station:	4			5			7			9		
	C-I	C/I	PO$_4$	C-I	C/I	PO$_4$	C-I	C/I	PO$_4$	C-I	C/I	PO$_4$
Depth (m) 0	36,173	6.6	0.045	18,840	4.1	0.102	−4,980	0.1	0.160	−5,040	0.3	0.030
50	26,073	5.1	0.010	15,340	3.5	0.045	32,020	6.9	0.060	34,960	6.0	0.010
100	−1,947	0.7	0.102	−3,160	0.5	0.025	−1,480	0.7	0.060	−2,040	0.7	0.066
Inoculum (cells/ml):	6,400			6,200			5,500			7,000		

LEO (3–6 August 1963)

Station:	1			2			3			4		
	C-I	C/I	PO$_4$	C-I	C/I	PO$_4$	C-I	C/I	PO$_4$	C-I	C/I	PO$_4$
0	−1,000	–	0.210	678,000	679.0	0.015	48,100	121.0	0.045	265,900	167.0	0.030
50	−1,000	–	0.595	166,500	167.5	0.295	11,100	29.0	0.030	157,900	100.0	0.030
100	−1,000	–	0.810	160,000	161.0	0.805	5,600	15.0	0.025	50,400	33.0	0.100
Inoculum (cells/ml):	1,000			1,000			400			1,600		

NEST (27 August–1 September 1964)

Station:	1			2			3			4		
	C-I	C/I	PO$_4$	C-I	C/I	PO$_4$	C-I	C/I	PO$_4$	C-I	C/I	PO$_4$
0	103,500	15.0	0.11	31,000	6.1	0.06	55,600	11.2	0	21,300	4.7	0
50	94,500	13.7	0.56	33,000	6.5	0.14	47,600	10.0	0.02	14,300	3.5	0
100	80,500	11.8	0.71	25,000	5.1	0.64	56,600	11.6	0.02	18,300	4.2	0.07
Inoculum (cells/ml):	7,400			6,000			5,400			5,700		

100 m water from all stations did not support growth. A 25 to 50% mortality of the inoculum concentration occurred during the five day incubation period. This contrasts with the 3.5- to 7-fold increase in inoculum found in unenriched 50 m water. Unenriched *surface* waters from stations 4 and 5 also supported good growth, but at stations 7 and 9 growth was suppressed to an even greater extent (70–90% inoculum mortality) than in 100 m water.

Re-examination of the *Cyclotella* response (Table 2) in the August 1963 (Cruise "Leo") bioassay experiments demonstrates that unenriched 100 m water was also generally less favorable for growth (Smayda, 1964). Complete inoculum lysis occurred at all depths at station 1 (Fig. 1). While active growth always occurred at stations 2–4, it became progressively poorer (relative to inoculum) with depth, similar to the response at Colombo stations 4 and 5.

In late August 1964 stations 1 to 4 were revisited during Cruise *Nest* (to be reported in detail elsewhere) to conduct additional *Cyclotella nana* assays. Unenriched 100 m water was again less favorable for growth than the upper layers at 2 of the 4 stations, though lysis did not occur (Table 2).

These three surveys demonstrate that unenriched 100 m water collected between 38° N and Bermuda during August 1963 and 1964, and between Bermuda and Puerto Rico during November 1964 is generally (8 of 12 cases) considerably less favorable for growth of *Cyclotella nana* (13-1) than 50 m water, i.e. an effect over a distance of about 20° of latitude! Further, two *surface* water provinces may be tentatively distinguished: one extending from stations 2 to 5, where active growth occurs and even generally exceeds that in 50 m water, and the other at stations 7 and 9 where growth is repressed. Examination of the data from these three cruises suggests that poorer growth of the unenriched samples relative to the inoculum generally coincided with higher initial phosphate levels (Table 2). Slight variations in technique between cruises, and local station-to-station variations in growth preclude direct comparison or rigorous analysis of this general, inverse relationship. The tendency of *unenriched* 100 m water to be generally less favorable to *Cyclotella* growth over a range of PO_4—P concentrations from 0.025 to 0.80 μg A L^{-1} further complicates this trend. Nonetheless, a more or less diffuse inverse relationship between growth in unenriched water and initial phosphate concentration is suggested by the data.

Growth in enriched water

1   *General responses*

Growth of *Cyclotella nana* (13-1) in various general enrichments during November 1963 ("Colombo") will be first compared to 1. the inoculum level; 2. to that in no enrichment, and 3. to that in complete enrichment (Table 3).

The response to BSM addition (i.e. P, N, Si + FeEDTA) shows similar trends when related to either the inoculum (I) level or to the unenriched (N) response. Significant growth occurred in 100 m water from all stations, and at station 7-0* relative to N. Poorest growth generally occurred in surface waters relative to I; it was generally equally poor in the upper 50 m relative to the unenriched response. Adding trace metals (BSM + TM) did not appreciably modify this general response, although growth was stimulated at 4-50, 5-50 and substantially reduced at 7-100 and, especially, 9-100.

Substituting vitamins (BSM + VITS) for trace metals considerably stimulated growth above that found with BSM, except at 5-100 and 7-100 where poor growth resulted (Table 3). Growth was generally similar in BSM with and without trace metals, which suggests that the response to vitamin enrichment was primarily to their addition and not to elimination of trace metals. At 9-100, however, a combination of these effects seems likely: trace metal addition significantly reduced growth, but vitamins were stimulatory. Vitamins were more inhibitory than trace metals at 7-100, and inhibitory at 5-100 where trace metals were without appreciable effect. Thus, vitamin addition stimulated growth at 4 of the 6 locations where unenriched water led to inoculum mortality, and inhibited growth twice relative to the beneficial influence of BSM when added to these waters.

Complete enrichment usually did not induce the best growth found for the general treatments, although it exceeded (maximum = 12, and based on I and N) that with BSM addition 10 times; 8 times with BSM + TM, but only twice† with BSM + VITS. The latter effect suggests that adding the trace metal mix to the BSM + VITS ration usually led to partial inhibition of growth.

These experiments indicate, therefore, that the BSM enrichment could remove the inoculum lysis always found in unenriched 100 m water (Table 2),

---

* This designation style will henceforth be used to identify station and depth. Thus, 7-0 refers to the surface at station 7.

† Three times when based on N.

Table 3 Growth of *Cyclotella nana* (13-1) in various enrichments relative to inoculum (I), no enrichment, i.e. None (N), and complete (ALL) enrichment (A)

Stations:	4			5			7			9		
Control:	I	N	A	I	N	A	I	N	A	I	N	A
BSM enrichment:												
0 m	2.3	0.4	0.04	5.3	1.3	0.15	5.0	55.0	0.07	1.2	4.3	0.05
50	12.1	2.4	0.56	3.1	0.9	0.19	8.8	1.3	1.62	22.9	3.8	0.27
100	39.1	55.6	46.0	20.6	42.6	0.68	40.8	56.0	0.55	34.2	48.0	0.53
BSM + TM:												
0	3.2	0.5	0.05	6.2	1.6	0.18	5.2	58.0	0.07	0.8	3.0	0.04
50	38.5	7.6	1.81	45.0	13.0	2.68	13.0	1.9	23.8	20.7	3.5	0.24
100	37.9	54.0	44.0	23.7	49.0	0.77	18.8	25.9	0.25	0.8	1.2	0.01
BSM + VITS:												
0	79.3	11.9	1.17	?	?	?	73.0	804.0	1.01	65.6	230.0	2.78
50	81.5	16.1	3.82	69.8	20.1	4.08	33.8	5.0	62.0	87.7	14.6	1.03
100	56.0	79.9	65.5	10.9	22.7	0.36	4.5	6.3	0.06	80.0	112.1	1.22
Complete Enrichment:												
0	67.5	10.2		34.9	8.7		72.1	794.0		23.5	82.5	
50	21.3	4.2		17.0	4.9		0.5	0.1		85.1	14.2	
100	0.8	1.2		30.5	63.4		72.1	102.0		65.2	91.5	

a treatment also leading to substantially greater growth than found in surface and 50 m waters. This treatment (relative to I) also overcame the inimical effect of the unenriched surface waters at stations 7 and 9, although growth was not stimulated to the same degree as in 100 m water. Vitamin addition (BSM + VITS) even further improved these waters for growth, except at 5-100 and 7-100. The addition of vitamins alone was not tested.

2   Effect of the BSM components

Eliminating the BSM components singly influenced growth (Table 4) and illustrates their relative importance in removing the inimical effects of the unenriched 100 m and surface waters. Their omission singly from 100 m water, unlike at other depths, always caused *poorer* growth (relative to I) than found in the unmodified BSM enrichment. The omission of nitrate at station 4, FeEDTA at stations 5 and 7, and nitrate and silicate at station 9 led to the poorest growth, or even inoculum mortality (stations 5, 9). For inimical surface water, at station 9 nitrate omission caused inoculum mortality (omitting FeEDTA markedly *stimulated* growth), whereas all omissions at station 7 stimulated or did not appreciably (PO_4^{-2})

Table 4   Growth of *Cyclotella nana* (13-1) without enrichment and with various combinations of P, N, Si and FeEDTA (BSM series) relative to inoculum

Station	Depth (m)	None	BSM	BSM			
				–P	–N	–Si	–FeEDTA
4	0	6.6	2.3	30.8	1.6	4.6	61.7
	50	5.0	12.1	10.9	5.5	1.2	13.2
	100	0.7	39.1	19.2	2.1	5.2	27.1
5	0	4.0	5.3	6.0	5.3	7.0	5.4
	50	3.4	3.1	9.6	11.2	9.4	6.6
	100	0.5	20.6	10.8	8.0	15.5	0.1
7	0	0.1	5.0	4.0	9.1	8.8	7.8
	50	6.8	8.8	3.1	7.0	6.4	1.7
	100	0.7	40.8	7.9	5.4	6.8	3.0
9	0	0.3	1.2	2.2	0.6	3.2	16.2
	50	6.0	22.9	10.0	3.2	2.2	23.5
	100	0.7	34.2	12.4	0.5	0.6	24.2

influence growth. This suggests that the factors responsible for inoculum lysis in the surface waters of stations 7 and 9 may differ from those in 100 m water.

The constancy and greater magnitude of the limiting effect of P, N, Si and FeEDTA on growth when omitted from 100 m water are conspicuous. Otherwise, their influence on growth and effectiveness in changing unenriched waters from "bad " to "good" waters are indistinguishable from that on waters initially "good", as at 50 m and the surface of stations 4 and 5. Thus, for 50 m water, silicate omission at stations 4 and 9 led to poorest growth (FeEDTA omission was without effect), as did FeEDTA at station 7, while all omissions *stimulated* growth at station 5. Nutrient omission was either stimulatory or without appreciable effect in the "good" surface waters.

The nutrient enrichments (Tables 3, 4) also reveal 1. the general tendency for a progressively greater, positive influence of BSM addition with depth at all stations, relative to either I or N as the control; 2. the opposite trend of the unenriched and BSM + VITS responses, leaving aside the "bad" surface waters of stations 7 and 9; and 3. the relative importance of omitting a given nutrient varied with depth. These trends are less apparent when the responses are related to growth in the complete enrichment, possibly due to the complications of the aforementioned growth reduction resulting from combining trace metals with vitamins.

3   *Growth relative to the BSM components*

The question of "good " and "bad " water for *Cyclotella nana* (13-1) growth, therefore, is linked somehow to the inorganic nutrient characteristics of these unenriched waters, as well as to physico-chemical properties influenced by chelation. The growth potential of 100 m water appears to be more dependent on inorganic nutrient enrichment (BSM) than the upper layers, which suggests that these nutrients are more limiting at this depth. These effects are further examined by comparing the growth in various enrichments to that in the BSM ration (Fig. 2) and to the inoculum (Table 5). BSM has been chosen as a control because this ration does not exhibit the observed complications of trace metal-vitamin interactions influencing response to complete enrichment; it always promoted growth, unlike NONE (ration 14); and inoculum lysis was always overcome by one or more of the ingredients (P, N, Si, FeEDTA) in this treatment.

The BSM ingredients (P, N, Si, FeEDTA) when omitted singly from 100 m water clearly limited growth relative to BSM (Fig. 2), and confirms the

Table 5   Growth of *Cyclotella nana* (13-1) in various modifications of the complete enrichment (ALL) relative to inoculum

Station	Depth (m)	ALL	-P	-N	-Si	-FeEDTA	-FeEDTA + Na$_2$EDTA	-TM	-VITS
4	0	67.5	17.6	4.5	10.5	2.8	33.8	79.3	3.2
	50	21.3	9.2	6.3	2.1	31.4	66.0	81.5	38.5
	100	0.8	23.9	6.4	3.9	35.3	39.2	56.0	37.9
5	0	34.9	11.5	11.0	25.3	6.6	22.7	?	6.2
	50	17.0	8.1	10.8	8.0	8.6	29.8	69.8	45.0
	100	30.5	1.8	10.6	24.5	14.1	24.0	10.9	23.7
7	0	72.1	10.0	6.1	9.9	13.6	23.0	73.0	5.2
	50	0.5	8.9	7.6	8.2	19.4	92.4	33.8	13.0
	100	74.1	1.1	6.7	6.1	68.3	5.1	4.5	18.8
9	0	23.5	5.7	1.4	1.9	35.3	37.5	65.6	0.8
	50	85.1	11.2	6.1	2.1	27.2	59.3	87.7	20.7
	100	65.2	0.7	3.4	0.7	1.1	106.4	80.0	0.8

effect noted previously using the inoculum as the control (Table 4). However, omitting FeEDTA from the upper 50 m, and phosphate and silicate (generally) from surface waters *stimulated* growth relative to that in BSM.

ENRICHMENT RESPONSE / BSM RESPONSE

M St.	0	50	100	0	50	100	0	50	100
	BSM − P			BSM − N			BSM − Si		
4	13.10	.90	.44	.66	.46	.05	1.96	.11	.13
5	1.15	3.09	.52	1.00	3.59	.39	1.32	3.00	.75
7	.80	.36	.19	1.84	.79	.13	1.76	.73	.17
9	1.88	.44	.36	.53	.14	.014	2.71	.10	.016
	BSM − FeEDTA			BSM + VITS			BSM + TM		
4	2.63	1.09	.70	33.85	6.74	1.41	1.37	3.18	.95
5	1.01	2.10	.007	n.d.	22.20	.53	1.18	14.30	1.15
7	2.73	2.20	.07	14.62	3.84	.11	1.05	1.47	.46
9	13.40	1.03	.71	54.10	3.82	2.33	.70	.94	.025
	ALL			ALL − P			ALL − N		
4	28.80	1.76	.02	.75	.76	.61	1.90	.52	.16
5	6.55	5.44	1.48	2.17	2.59	.09	2.79	3.44	.51
7	14.40	.06	1.82	2.00	1.01	.027	1.27	.86	.16
9	19.40	3.71	1.90	4.76	.49	.023	1.18	.27	.10
	ALL − Si			ALL − FeEDTA			ALL − Fe (+ Na_2EDTA)		
4	.44	.18	.10	1.20	2.60	.90	14.40	54.50	1.01
5	4.76	2.60	1.19	1.26	2.74	.69	4.28	9.50	1.16
7	2.00	.94	.15	1.56	.19	16.80	4.60	10.50	1.27
9	1.59	.09	.02	29.20	1.18	.033	31.00	2.58	3.50

Figure 2 The response of *Cyclotella nana* (clone 13-1) to various nutrient treatments relative to that in the BSM enrichment (i.e. P, N, Si + FeEDTA) during *Colombo*. The nutrient emendment BSM-P indicates phosphate was omitted from the BSM ratio. ALL represents the complete enrichment (ration 8)

That is, adding these nutrients *inhibited* the potential growth of *Cyclotella nana* (13-1). Nitrate omission, however, generally limited growth. A general, progressive decrease in ratio of growth relative to control generally occurred with depth for all nutrient omissions at all stations.

The omission of P, N, Si and FeEDTA singly from the *complete* enrichment (Fig. 2) generally had a similar effect on growth as that found in the equivalent BSM modifications. Significant growth promotion accompanied vitamin addition (BSM + VITS), except in 100 m water at stations 5 and 7. This effect progressively decreased with depth and confirms the response trend found previously using inoculum as the control (Table 3). The addition of the trace metal mix (BSM + TM) inhibited growth at station 9 and 7-100 in contrast to its stimulatory effect elsewhere.

The omission of Fe from the complete medium, but with EDTA added as 1 mg/L in Na^+ form (ration 13), usually greatly stimulated growth in the upper 50 m except at station 9, whereas 100 m water was less affected by this manipulation. Possibly, Fe addition to these upper waters was unfavorable for *Cyclotella* growth (as also suggested by the relative response to BSM-FeEDTA) whereas adding Na_2EDTA improved water quality. The specific effects of Fe and EDTA presence on growth are difficult to sift out, aside from clear-cut evidence that 100 m water was not similarly influenced as the upper layers.

These nutrient effects are generally confirmed by the responses of complete enrichment modifications relative to inoculum (Table 5), although the greater opportunity for synergistic effects and station-to-station variations in growth conditions complicate interpretation. But, in general, vitamin addition to surface waters is generally more effective in promoting growth than when added to deep water, and the omission of N, P, Si and FeEDTA generally limited growth, especially in 100 m water.

DISCUSSION

The *Colombo* results and the August 1963 bioassay study conducted about three months earlier generally agree (Smayda, 1964). The main conclusions of the latter study are presented in the introduction. The most interesting new finding is the demonstration that *unenriched* 100 m water was usually "bad" for *Cyclotella nana* (13-1) during August 1963, 1964 and November 1964, and led to inoculum lysis in the present experiments. ("Bad" surface water was also encountered between Bermuda and Puerto Rico (stations 7, 9), but the data are insufficient to consider this of equally persistent occurrence as the 100 m effect.) This effect extended over 20° of latitude, or roughly from New England slope water to Puerto Rico (Fig. 1), which suggests it to be a "water quality" characteristic of regional importance. The inimical effect was clearly removed by enriching 100 m water with phosphate, nitrate, silicate and FeEDTA in various combinations. Growth was usually stimulated even further by vitamin addition, but not by trace metals. In contrast, adding the major inorganic nutrients to 50 m and, especially, surface waters, often inhibited *Cyclotella* growth. There vitamin addition was particularly stimulatory, whereas this effect was considerably reduced in 100 m water. Despite these two general trends, the fundamental causes for the initial occurrence of "bad" 100 m water and "good" 50 m water (both types found at surface) remain unknown. Presumably the 100 m phenomenon is not an experimental artifact, although the problems associated with nutrient enrichment techniques, including nutrient overloading and balance, precipitation, chelation and time-effects can not be ignored. Tranter and Newell (1960) have reported a mortality of *natural* populations over a 20 hour period when enclosed in 500 ml translucent polyethylene bottles which could be spared usually by $FeSO_4$-EDTA enrichment, but not by nitrate or phosphate. However, the ensuing discussion will stress the *Cyclotella nana* (13-1) response, and general ecological extrapolation to growth conditions or natural populations in these watermasses is not to be inferred.

An unequivocal hydrographic separation (aside from depth) between "bad" and "good" unenriched waters is not possible (Table 1). The very general, diffuse inverse relationship between growth in the unenriched assay (relative to inoculum) and phosphate concentration suggested by the data (Table 2) is too tenuous to warrant serious discussion. This indicates that an effect beyond that detectable through routine hydrographic measurements is operative. However, a curvilinear relationship can be established between growth in the BSM enrichment (P, N, Si, FeEDTA), relative to the inoculum, and phosphate concentration when plotted on semi-log paper (Fig. 3A). Ignoring the "aberrant" data point for station 7-50, two separate curves result: 1. a lower curve which includes all the surface samples and 5-50, and 2. an upper one comprising all 100 m water and stations 4-50 and 9-50. Growth progressively increased with phosphate concentrations up to about 0.06 to 0.08 μg A L^{-1} and remained constant above this "saturation" value. However, for a given phosphate concentration significantly greater growth accompanied BSM enrichment of 100 m water than surface water. It was about 8-fold greater at the asymptotic responses above 0.06 μg A PO$_4$—P L^{-1}. The BSM enrichment contained 5 μg A PO$_4$—P L^{-1} which considerably exceeds the maximum concentration (0.16 μg A L^{-1}) reported for the upper 100 m near station 4 (Menzel and Ryther, 1960). The anti-

13*

cipated range in available phosphate (Table 1) from about 5.01 to 5.16 μg
A L^{-1}, therefore, would appear to be insignificant. Presumably the added
nitrate, silicate and FeEDTA concentrations (see ration 1) were also neither
limiting, nor caused significant differences between samples (Table 1),
considering the measured concentrations of these nutrients (Menzel and
Ryther, 1960; Menzel and Spaeth, 1962a, b). Nonetheless, initially low
phosphate waters (below ∼0.08 μg A L^{-1}) are demonstrably less receptive
to growth stimulation by BSM addition than phosphate richer waters,
independent of the depth effect. Neither this result nor the unexpected deep
water response can be accounted for, including after examination of the
Cyclotella response to other nutrient enrichment treatments (Tables 2–5)
and also graphing many of these responses against phosphate concentration.
It seems probable that phosphate *per se* is not the important factor, but some
presently unidentified component(s) whose effect runs parallel to the phosphate concentration as defined in Figure 3A.

A similar curvilinear relationship characterizes the response to silicate
omission from BSM (relative to the inoculum) when related to phosphate
concentration (Fig. 3B). The 100 m responses at stations 5 and 9 clearly
deviate from this general trend which is independent of depth unlike that
with BSM (Fig. 3A). An increase in response to silicate omission occurs
up to the phosphate "saturation" value of about 0.06 μg A L^{-1}, above
which growth is constant. Silicate values are lacking for station 9 (Table 1),
but plotting the BSM-Si response at the other stations against silicate concentrations reveals no relationship. Thus silicate omission appears to limit
growth in phosphate-poor waters in an inexplicable way.

The thermal structure at station 4 (Fig. 1, Table 1) suggests that the
seasonal thermocline (Menzel and Ryther, 1960) was beginning to break
down. Perhaps, then, 100 m water here is "newer" than the shallower
layers. There is evidence that "new" (or "deep") water requires a "sea-
soning", despite high inorganic nutrient levels, to permit phytoplankton
growth. Chelators and/or natural organic extracts have been implicated
(Menzel, Hulburt and Ryther, 1963; Barber and Ryther, 1969). However
significant growth improvement always occurred after phosphate, nitrate
and silicate addition, *without* FeEDTA (Table 4). If this is a "new" (100 m)
versus "old" (surface) water problem at station 4, then improvement of
both water types for *Cyclotella nana* (13-1) is not fundamentally a problem
of chelation. However, the widespread persistence of the 100 m response
(and possibly the surface response at stations 7 and 9) included stations

where the seasonal thermocline is firmly established (Table 1). This suggests that the question of "bad" and "good" water in my experiments is not related to the degree of mixing or "newness" of water. The greater stimulation of *Cyclotella* growth by vitamin addition to surface waters and the

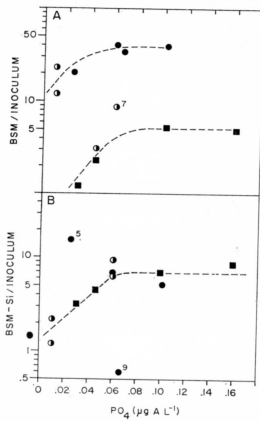

Figure 3A The relationship between growth of *Cyclotella nana* (clone 13-1) in the BSM enrichment relative to the inoculum and the *in situ* phosphate concentration. **3B** This relationship omitting silicate from the BSM enrichment. Curves drawn by eye. Symbols: ■ surface; ◑ 50 m, and ● 100 m samples. Numbers, where given, refer to stations

progressive decrease in its effect with depth (Fig. 2) are likewise not explicable from the known vitamin concentrations characterizing these waters (Menzel and Spaeth, 1962c), or the general bioassay and hydrographic observations.

SUMMARY

1. The growth of *Cyclotella nan a* (clone 13-1), incubated at 600 foot candles and 20 C for five days, in 168 enrichment experiments (14 rations) using filtered water collected from the surface, 50 and 100 m at four stations between Bermuda and Puerto Rico during November 1963 is presented.

2. The occurrence of "bad" water at 100 m, when unenriched, at all stations which caused inoculum mortality is demonstrated. This "water quality" characteristic has been found on two other cruises, and extends over 20° of latitude.

3. The addition of phosphate, nitrate, silicate and FeEDTA (BSM enrichment) to 100 m water always removed the inimical effect; growth was generally stimulated even further by vitamin addition, but not by trace metals.

4. Adding the major inorganic nutrients to 50 m and, especially, surface waters often inhibited growth relative to the control, whereas vitamin addition was particularly stimulatory in contrast to its influence on 100 m water.

5. Nitrate and silicate omission frequently limited growth.

6. Curvilinear relationships between growth in the BSM enrichment, with and without silicate, and phosphate concentrations are found.

7. The conditions initially responsible for the unenriched, 100 m ("bad") water effect on *Cyclotella nana* (13-1) growth and the 50 m "good" water effect (both types found at surface) remain unknown.

Acknowledgement

The assistance of Miss Brenda J. Boleyn during all phases of the investigation is acknowledged. Dr. Robert R. L. Guillard provided the culture of *Cyclotella nana* (13-1). This investigation was aided partly by grant GB-5366 from the National Science Foundation.

References

BARBER, R. T. and RYTHER, J. H. (1969). Organic chelators: factors affecting primary production in the Cromwell Current upwelling. *J. exp. mar. Biol. Ecol.* **3**, 191–199.

GUILLARD, R. R. L. and RYTHER, J. H. (1962). Studies of marine planktonic diatoms. I. *Cyclotella nana* Hustedt and *Detonula confervacea* (Cleve) Gran. *Can. J. Microbiol.* **8**, 229–239.

JOHNSTON, R. (1963). Sea water, the natural medium of phytoplankton. I. General features. *J. Mar. Biol. Ass. U. K.* **43**, 427–456.

JOHNSTON, R. (1964). Sea water, the natural medium of phytoplankton. II. Trace metals and chelation, and general discussion. *J. Mar. Biol. Ass. U. K.* **44**, 87–110.

MENZEL, D. W., HULBURT, E. M. and RYTHER, J. H. (1963). The effects of enriching Sargasso Sea water on the production and species composition of the phytoplankton. *Deep-Sea Res.* **10**, 209–219.

MENZEL, D. W. and RYTHER, J. H. (1960). The annual cycle of primary production in the Sargasso Sea off Bermuda. *Deep-Sea Res.* **6**, 351–367.

MENZEL, D. W. and SPAETH, J. P. (1962a). Occurrence of ammonia in Sargasso Sea waters and in rain water at Bermuda. *Limnol. Oceanogr.* **7**, 159–162.

MENZEL, D. W. and SPAETH, J. P. (1962b). Occurrence of iron in the Sargasso Sea off Bermuda. *Limnol. Oceanogr.* **7**, 155–158.

MENZEL, D. W. and SPAETH, J. P. (1962c). Occurrence of vitamin B_{12} in the Sargasso Sea. *Limnol. Oceanogr.* **7**, 151–154.

SMAYDA, T. J. (1964). Enrichment experiments using the marine centric diatom *Cyclotella nana* (clone 13-1) as an assay organism. *Proc. Symp. Mar. Ecology, Occ. Publ. Grad. School Oceanogr. Univ. Rhode Island* No. 2, 25–32.

SMAYDA, T. J. (1970). Growth potential bioassay of water masses using diatom cultures: Phosphorescent Bay (Puerto Rico) and Caribbean waters. *Helgoländer wiss. Meeresunters* **20**, 172-194.

TRANTER, D. J. and NEWELL, B. S. (1963). Enrichment experiments in the Indian Ocean. *Deep-Sea Res.* **10**, 1–9.

Mesoscale studies of the physical oceanography in two coastal upwelling regions: Oregon and Peru

ROBERT L. SMITH,* CHRISTOPHER N. K. MOOERS,†
and DAVID B. ENFIELD

*Department of Oceanography, Oregon State University
Corvallis, Oregon U.S.A.*

Abstract

The emphasis in this paper is on the physical process of coastal upwelling, the process by which water from depths of up to a few hundred meters offshore is introduced into the upper few meters, and the euphotic zone, near the coast. Upwelling is in general the result of horizontal divergence in the surface layer and is a particularly conspicuous phenomenon along the western coasts of the continents in middle and low latitudes, where seasonal equatorward winds produce both an equatorward and offshore flow in the surface layer because of the earth's rotation.

Arrays of recording current meters and thermographs have been moored on the continental shelf off Oregon (USA) during the upwelling season (northern summer) since 1965. These measurements have been supplemented by extensive hydrographic measurements and, more recently, airborne infra-red thermometry. The general flow pattern can be characterized by: 1. Equatorward flow of about 20 cm/sec occurs in the upper 40 meters, with an offshore component in the surface Ekman layer about

---

* Present address: Office of Naval Research, Ocean Science and Technology Division, Washington, D. C.

† Present address: Rosenstiel School of Marine and Atmospheric Sciences, University of Miami, Miami, Florida.

20 meters thick. 2. An undercurrent, or poleward flow, of about 10 cm/sec occurs beneath the inclined permanent pycnocline with an onshore component in the upper pycnocline and beneath the pycnocline. 3. The cold, saline water upwelled from beneath the permanent pycnocline is modified by mixing and absorption of heat while near the surface, and flows offshore sinking along the base of the permanent pycnocline. This modified water is warmer (but more saline) and appears as a temperature inversion.

Similar measurements were made off Peru at 15° S, where upwelling was occurring close to the coast, in the early southern autumn of 1969. The density stratification was small and the current field beneath a shallow surface Ekman layer was quasibarotropic with a mean speed of about 20 cm/sec. The mean flow beneath the surface layer was poleward (to the southeast), against the prevailing winds. There was an onshore component to the mean flow. A large scale barotropic 'event', unrelated to local meteorological conditions, caused the currents to reverse and flow equatorward for four days. No significant change in the upwelling was apparent. It is suggested that the mean poleward flow is a manifestation of the Peru–Chile Undercurrent over the continental shelf.

INTRODUCTION

Coastal upwelling is a process in which water from depths of up to a few hundred meters is brought into the upper several meters near shore. Coastal upwelling is at certain seasons the predominant mesoscale physical process along the low- and mid-latitude western coasts of most continents. Because of the earth's rotation, equatorward winds along these coasts produce both equatorward and offshore flow in the "surface" layer necessitating an upwelling of "subsurface" water near the coast. Biologically, the coastal upwelling regions are among the most important regions in the ocean. The primary production is very high and, as Ryther (1969) has discussed, the shortness of the food chain in these regions may enable half of the world's fish supply to the produced there.

As important as the upwelling process is, Wooster (U. S. Coast Guard, 1969) has pointed out: "We are still guessing about the speed of vertical motion in an upwelling area. We have numbers and methods which are about the same as those used by George McEwen back in the Twenties and early Thirties of this century". In a review article Smith (1968) discusses the work on upwelling to 1967. Compared with studies of the western boundary currents there has been relatively little theoretical and observational work done in physical oceanography on coastal upwelling, or on eastern boundary currents in general.

The qualitative explanation of coastal upwelling is based on Ekman's classic work (1905) which showed that in a homogeneous, infinite ocean, a uniform steady wind would produce a net transport of water τ/f in a direction 90° *cum sole* to the wind. (τ is the wind stress and f the Coriolis parameter.) The component of the Ekman transport directed offshore, computed from the mean wind stress charts, does give a rather good prediction of where and at what season coastal upwelling occurs. The Ekman transport has been used in analyzing hydrographic data in the upwelling region of the California current system by McEwen (1912) and Sverdrup (1938). In some mean or climatological sense we understand—or can explain—upwelling.

But the time-series observations made by today's oceanographers clearly show that neither the wind nor current field is either steady or statistically stationary; variability and intermittency are predominant. In this paper we will discuss some intensive, mesoscale, time-series studies of the circulation in two of the major upwelling regions: off Oregon and off Peru. These are indicative of one of the types of studies that is necessary for the physical oceanographer to do if we are to understand upwelling, i.e. to have a predictive knowledge and to be able to relate the measurements of the physical oceanographer and the biological oceanographer.

THE OREGON COASTAL UPWELLING REGIME

Upwelling is a seasonal phenomenon in the California Current region, reflecting the seasonal changes in the position and strength of the North Pacific High and the Aleutian Low pressure systems. Off Oregon, the North Pacific High dominates the atmospheric circulation from spring to early fall. The winds are predominantly northerly or northwesterly along the Oregon coast, and coastal upwelling occurs.

Figure 1 shows the sea surface temperature on 12 August 1969. While the picture may be considered typical of coastal upwelling, the synopticity and detail of the map is not typical. The data was obtained from an airborne precision radiation thermometer (ART) accurate to 0.1°C, and the track shown was flown in less than 6 hours. A series of flights with the ART has provided the first semi-synoptic picture of coastal upwelling along the Oregon coast. While upwelling was apparent on all the flights in July and August, the intensity of upwelling (as measured by temperature) and sharpness of the horizontal temperature gradients changed markedly in a matter of days.

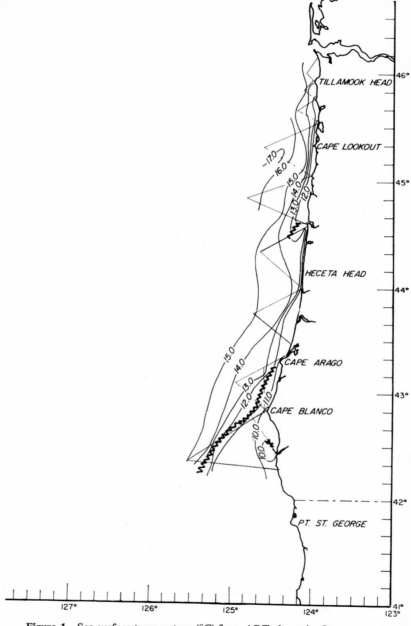

Figure 1 Sea surface temperature (°C) from ART along the Oregon coast, 12 August 1969. Heavy jagged lines indicate color discontinuity

The pronounced upwelling south of Cape Blanco is readily apparent in Figure 1. The intensification of upwelling equatorward of capes in eastern boundary currents has been demonstrated theoretically by Arthur (1965) and Yoshida (1967).

Across the intense temperature gradient or front south of Cape Blanco, the temperature changed 2.5°C in 10 km. Fronts are a manifestation of the internal adjustment of the density field in the coastal waters. During the upwelling season the permanent pycnocline or frontal layer, which can be considered to be bounded by the 25.5 and 26.0 sigma-t surfaces, slopes up from 100 m depth 50 km offshore and may intersect the surface 10 to 20 km offshore forming a surface front.

Figure 2 shows the vertical sections of sigma-t, temperature and salinity made off Oregon at about 45° N in August 1966. They may be considered typical of the Oregon coast during upwelling, although the depth and intensity of the front may vary. The seasonal pycnocline is the result of both a seasonal halocline resulting from the influence of the Columbia River plume and the seasonal thermocline. The permanent pycnocline was the result of the permanent halocline, and in fact, the temperature inversion near its base weakened the effect of the halocline. The temperature inversion is often associated with the frontal layer—and is most intense inshore and washes out as it follows the descent of the frontal layer offshore. The inclined frontal layer gives rise to strong baroclinicty. If the interior is in 'quasi-geostrophic' equilibrium we might expect what the meteorologists call a 'thermal wind'. A vertical shear in geostrophic current is associated with a horizontal gradient in the density field.

Since 1965 we have been mooring arrays of recording current meters (BRAINCON-316) and thermographs (BRAINCON-146) on the continental shelf off Oregon during upwelling. In August and September 1966 meters were moored as indicated in Figure 2. The intention was to have current meters in the surface layer, the frontal layer and below the frontal layer (the lower layer). The position of the frontal layer is not stationary and thus the vertical position of a moored current meter relative to the frontal layer may vary. The moorings were made at DB-5, DB-10, and DB-15, which were 10, 20 and 30 km offshore in water depths of 80, 140 and 200 m respectively. The mooring technique is described in Pillsbury *et al.* (1969).

A convenient graphical view of the mean flow, and periodic patterns, and non-periodic variability in the currents is given by progressive vector diagrams (PVD). A PVD is constructed by the vector addition of successive

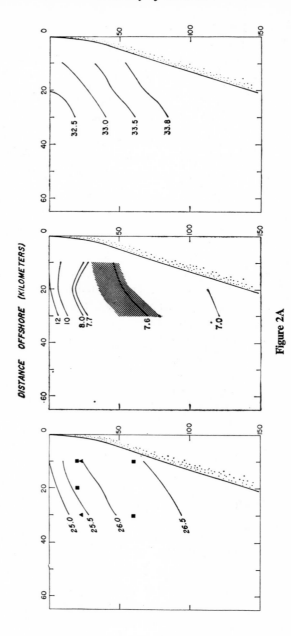

Figure 2A

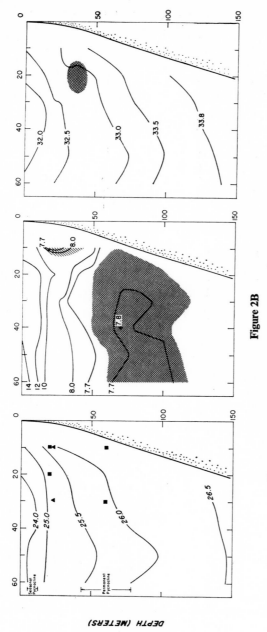

Figure 2B

Figure 2 Sigma-t, temperature and salinity off Oregon (45° N) on
A) 15–16 August 1966
B) 26–29 August 1966

Solid square indicates moored current meter
Solid triangle indicates moored thermograph
Shaded area indicates temperature inversion

DEPTH (METERS)

horizontal velocity measurements at a fixed point—and is not generally the trajectory of a water parcel. Figure 3 shows the PVDs for the entire period of the installation, 15 August to 24 September 1966. These PVDs demonstrate that the flow was to the south in the surface layer and to the north in the lower layer.

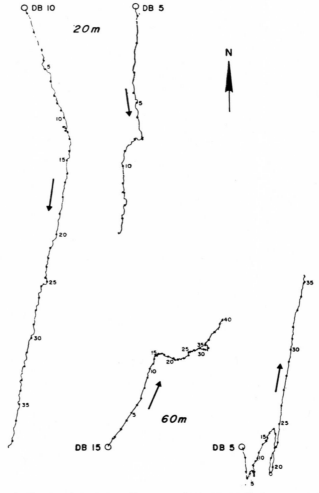

Figure 3 Progressive vector diagrams of currents off Oregon (45° N), 15 August to 24 September 1966. Midnight of each day denoted by closed circle

The winds during the first three weeks of August 1966 were strong from the north-northwest, which was the favorable condition for upwelling. The last week of August had weak and variable winds, often from the south-west—an unfavorable condition for coastal upwelling.

In Table 1 the statistics for the mean flow are given for the 14 day period (15 to 30 August 1969). The data is based on 339 hourly averages of the 10-minute integrated measurements recorded by the instruments. The mean

Table 1 Summary of mean flow statistics off Oregon: 15 to 30 August 1966 (means and standard deviations in cm/sec)

Station	Sensor depth	Northward component	Eastward component	Scalar speed	Alongshore (poleward)	Onshore
DB-5	20 m	−17.9 ± 12	−2.2 ± 11	23.4 ± 7	−16.6	+7.1
DB-5	60 m	+0.7 ± 9	+2.8 ± 6	11.0 ± 3	+2.0	+2.0
DB-10	20 m	−12.1 ± 9	+4.5 ± 10	18.1 ± 6	−8.3	+9.9
DB-15	60 m	+7.9 ± 13	+6.0 ± 6	13.7 ± 7	+9.8	+1.2

temperature at 20 m, 10 km off the coast was 8.4°C ± 0.6, while at 30 km off the coast it was 10.2°C ± 1.2. Since the bottom topography is not precisely north-south, but lies along 020° to 200° T, the onshore and longshore components are also given. All meters indicate a mean onshore flow. The near-surface offshore flow occurs in the surface Ekman layer where the frictional effects dominate. The surface Ekman, or wind drift layer must be less than 20 m in depth. It will be shown that the wind-drift layer off Peru is also shallow. More complete and extensive discussions of the measurements off Oregon can be found in Collins, *et al.* (1968) and Mooers (1969).

Based on the series of mesoscale investigations of the circulation in the coastal upwelling region off Oregon, we are able to present a generalized simple model of coastal upwelling in that relatively highly stratified part of the ocean.

A pictorial description of a simple model for the steady-state coastal upwelling process is given in Figure 4. A verbal description of the coastal upwelling flow regime follows (Mooers, 1969):

1 In the summer season, the mean north-northwesterly winds cause a net offshore transport of water in the surface Ekman layer, which is of the order of 10 meters thick. They also cause a southward flow in the alongshore direction, which can be thought of as the barotropic component of the alongshore flow, to a depth of about 40 m.

2 Mass compensation requires a net onshore transport of water below the surface Ekman layer. The flow is onshore in the upper portion of the permanent pycnocline, and presumably in a bottom Ekman layer, 10 to 20 m thick.

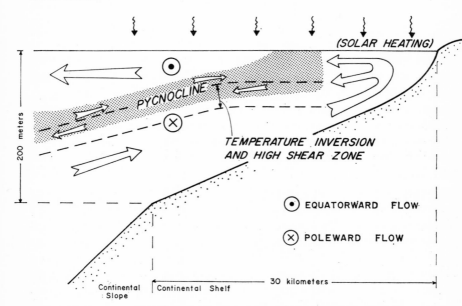

Figure 4 Schematic of mean circulation during upwelling off Oregon

3 Since the water of the open ocean reservoir is density stratified with a permanent pycnocline at a depth of about 100 meters, the net offshore transport of light water near the surface and the net onshore transport of heavy water near the bottom cause the permanent pycnocline to rise inshore, forming an inclined frontal layer.

4 The inclined frontal layer induces a "thermal wind". Thus, there is a baroclinic component to the alongshore flow such that the flow in the inclined frontal layer is increasingly poleward relative to the surface flow as the depth increased. To the extent that the barotropic and baroclinic flow components can be considered linearly superimposed, the barotropic flow

induced by the wind, or inclination of the sea surface, either operates to rein-
force or to cancel the baroclinic flow in the frontal layer. For instance, with the
permanent pycnocline at a fixed intensity and inclination, if the barotropic flow
is sufficiently southerly, the lower layer can come to a standstill or be reversed
to equatorward flow, while the surface layer continues to flow equatorward.
On the other hand, if the barotropic component relaxes, or reverses, the
upper layer can come to a standstill or be reversed to poleward flow, while
the lower layer continues to flow poleward. Reversals similar to those
described are observed to occur on a time scale of several days to
weeks.

5 The inclined frontal layer is a region where the processes of fronto-
genesis, necessary for the development and the sustenance of the frontal
layer, and of frontolysis, necessary for the destruction of the frontal layer,
are of significance. Appreciable mixing occurs in the frontal zone. If mixing
is sufficiently intense, the isopycnals in the layer beneath the inclined frontal
layer become downwarped, intensifying the tendency for northward flow
there. The mixing of warm, fresh waters derived from the surface layer with
cold, saline waters upwelled from the lower layer near the surface front
replenishes the water mass of the frontal layer. The freshly formed water
mass of the frontal layer sinks to the lower half of the inclined frontal layer
and below and then flows seaward, adding to the volume of the permanent
pycnocline. The modified, or mixed, upwelled water is warmer, but more
saline, than water of similar densities offshore and thus appears as a temperature inversion. The temperature inversion water presumably 'leaked'
through the current meter array, due to the large separation between current
meters. The hydrological-optical investigation of Pak *et al.* (1970) supports
the deduced cross-stream flow pattern.

6 The seasonal pycnocline develops at the base of the surface Ekman
layer. It is formed by the seasonal thermocline, which develops from summer
heating, and by the seasonal halocline, which is derived from the mixing
of surface layer water with the relatively fresh water of the Columbia River
plume. The seasonal pycnocline breaks the sea surface to form a surface
front, which tends to block the offshore flow of lower layer water that has
been supplied from the upper portion of the permanent pycnocline to the
surface Ekman layer inshore.

7 The above remarks present only a steady-state model. When the winds
accelerate sufficiently, the permanent pycnocline becomes more steeply in-

14*

Fertility of the Sea

clined and breaks the surface, forming a surface front, while the surface front formed by the seasonal pycnocline propagates offshore, causing a strong surface divergence to develop. The acceleration process may be largely advective. When the winds decelerate, the response is less rapid because the process of developing an inclined frontal layer is essentially irreversible on the short time scale, requiring mixing for its destruction.

A STUDY OF A PERU COASTAL UPWELLING REGION

In late March and April 1969 an intensive study was made of coastal up-welling along the Peru coast near 15° S. Scientists from several nations and all disciplines of oceanography participated. The analyses are not yet finished but we will discuss here the preliminary physical oceanography results obtained during the first two weeks, the survey phase, of the study.

The study was made in the early southern fall and not during the season of most intense upwelling off Peru, the southern winter. However, the available sea surface temperature charts (e.g. Eber *et al.* 1968) showed a persistent center of upwelling near 15° S. A survey grid approximately 100 km long and extending about 30 km out from the coast was established. Grid patterns were run by the R/V THOMPSON at full speed using an underway data acquisition system, which sampled water from 3 meters depth continuously for temperature, nitrate, silicate and chlorophyll 'a'. The approach was similar to that of Armstrong *et al.* (1967) but with the additions of a digital data logging capability and a shipboard computer which allowed maps of surface properties to be made rapidly. Upon first reaching the area, a grid pattern was run and the existence of upwelling in the region near Cabo Nazca confirmed.

Figure 5 shows the general area chosen for study and the sea surface (3 m) temperature obtained on grid patterns run on 27/28 and 28/29 March. The general shape of the contours of nitrate, silicate, and chlorophyll 'a' are very similar to that of temperature. Associated with the coldest temperature (16°C or less) were nitrate values greater than 18 μg-atom/l, silicates ranging from 10 to more than 20 μg-atom/l, and low chlorophyll 'a' (usually less than 2 μg/l). In the region of high temperature (19°C or greater), nitrates and silicates were less than 2 μg-atom/l. Chlorophyll 'a' varied considerably out of the coldest temperatures and in the pattern shown reached 8 μg/l near CM-2. Various grid patterns were repeated throughout the survey, allowing a quasi-synoptic picture of the surface manifestation of the up-

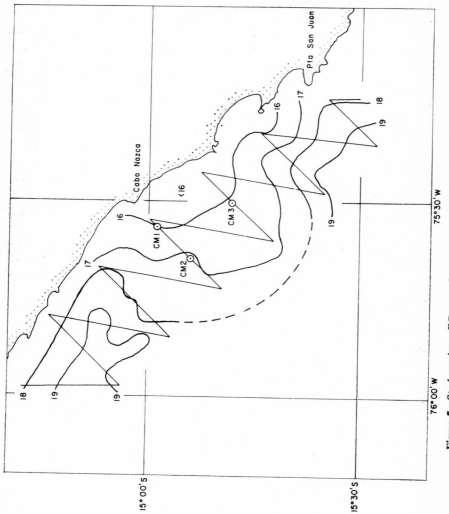

Figure 5   Study region off Peru and sea surface temperature, 27–29 March 1969

Fertility of the Sea

welling. During the survey phase (27 March to 10 April) six grid patterns were obtained that encompassed sufficient area to indicate the upwelling center. The gross pattern persisted throughout the two weeks and a center of cold water (temperature less than 16°C) with high nutrients (nitrate values about 20 μg-atom/l, silicate greater than 10 μg-atom/l) and low chlorophyll 'a' (less than 2 μg/l) could always be identified somewhere near the coast between Cabo Nazca and Punta San Juan. Away from the upwelling center the details varied, especially the nutrient ratios and position of chlorophyll 'a' maxima. The variations provide fertile material for the biologists and we shall leave the interpretation to them—noting only that there was a persistent upwelling source or sources throughout the period.

Three strings of recording current meters were moored at the sites indicated in Figure 5 as CM-1, CM-2 and CM-3. A recording anemometer was attached to the surface buoy at CM-2. CM-1, moored about 10 km offshore in 80 meters of water, and CM-2, moored about 20 km offshore in 180 m of water, were successfully recovered. The measurements were made and recorded over successive 20-minute periods for approximately 12 days. The wind during the entire period was from the southeast quadrant, essentially parallel to the coast. The mean wind speed was 9 ± 3 knots, and varied from a calm to a maximum speed recorded by the anemometer of 15 knots. A diurnal modulation in the speed was apparent. The depths at which current meters were moored and the mean statistics for the current meter records are given in Table 2. These data will be discussed in more detail below.

Two hydrographic sections, extending from inshore at Cabo Nazca and from Punta San Juan to about 100 km off the coast, were made. The typical upward slope of the isolines of physical and chemical properties from 50 km

Table 2 Summary of mean flow statistics off Peru: 28 March to 10 April 1969 (means and standard deviations in cm/sec)

Station	Sensor depth	Alongshore (to NW) component	Onshore (to NE) component	Scalar speed	Record length (hrs)
CM-1	25 m	−8.6 ± 15	+3.2 ± 7	17.6 ± 7	286
CM-1	50 m	−15.4 ± 18	+0.7 ± 6	22.5 ± 11	234
CM-2	25 m	−17.9 ± 17	+4.2 ± 7	23.7 ± 11	282
CM-2	50 m	−15.2 ± 16	+4.6 ± 8	22.5 ± 9	277
CM-2	135 m	−6.6 ± 9	−0.7 ± 3	10.4 ± 6	180

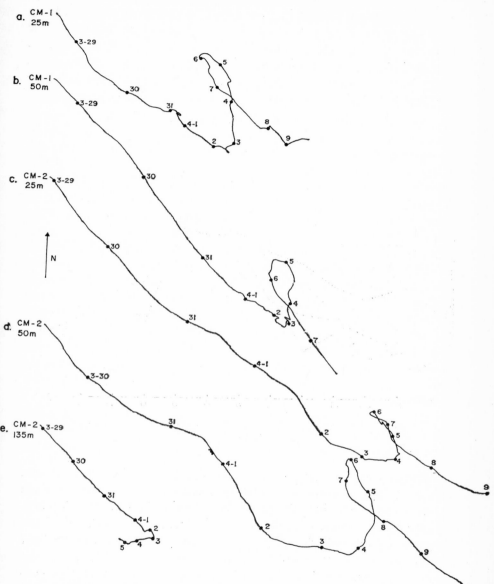

Figure 6 Progressive vector diagrams of currents off Peru, 20 March to 10 April 1969

offshore appeared. The depths from which the isolines began to slope up were 75 m or less, compared to 150–200 m off Oregon. Both the slopes of the isopycnals and the stability (density stratification) were less than off Oregon. The stability was about 1/3 that off Oregon. Thus the 'thermal wind' effect, so important off Oregon, would be considerably reduced and considerably less velocity shear might be expected.

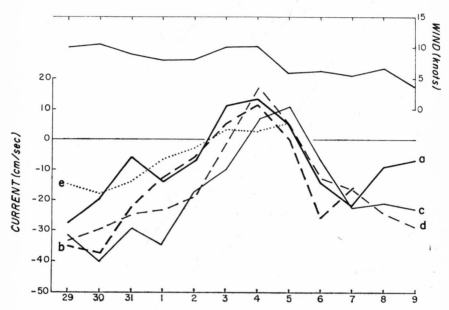

Figure 7   Daily mean wind and longshore currents for meters identified in Figure 6

Figure 6 shows the progressive vector diagrams for the current meters. The flow patterns are similar at all depths, but there are two surprises: 1. the mean current over the period, as shown in the PVDs and Table 2, is toward the South, opposite to the wind; 2. an 'event', the several day gradual reversal of the currents and then back, occurs. In Figure 7 is plotted the daily mean longshore component of the current for each sensor and the daily mean wind speed. Since the wind blew consistently from the southeast, parallel to the coast, the mean speed would not differ appreciably from the mean longshore wind. The current, and the 'event', are essentially independent of depth and hence 'quasi-barotropic'.

The wind-drift layer is apparently less than 25 meters deep. There is nothing in the wind record that would explain the reversal of the currents.

To study the near surface currents, drogues with a large current cross at 10 m were launched on 8 occasions and tracked from 2 to 28 hours. The drogues, or 10 m currents, were compared with the 25 m current meter records and winds for the appropriate times. When the winds were greater than 10 knots there was component of the 10 m current in the direction of the wind, but with weaker winds the drogues moved against the wind.

In Figure 8, the vectors representing the wind, the net movement of the drogue, and the vector mean of the nearby 25 m current record during the drogue tracking, are shown for 2 cases. Drogue 5 was tracked for 6 hours

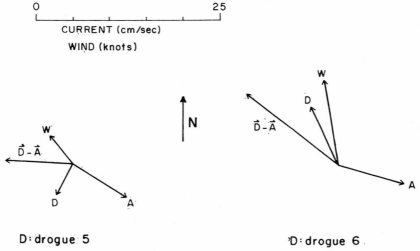

D: drogue 5   D: drogue 6

Figure 8 W: wind vector; A: current vector at 25 m; D: 10 m drogue
trajectory; D–A: wind-drift current vector at 10 m

during winds of 5 knots, and the drogue moved against the wind. Drogue 6 was tracked for 8 hours during winds of 12 knots, and the drogue moved with the wind. Since the current meters indicated a 'quasi-barotropic' flow, the vector difference between the 10 m (drogue) current and the 25 m current is indicative of the wind-drift current. In general there was a relative velocity shear in the sense one would expect from Ekman theory. In Figure 8 the wind-drift currents are in a direction about 45° to the left of wind.

The mean flow to the south may be evidence of the Peru–Chile Undercurrent discussed by Wooster and Gilmartin (1961) and Wyrtki (1963). This would indicate the undercurrent is present to the surface over the continental slope and shelf at times. The current reversal or 'event' appears to be a displacement of the undercurrent offshore by an intrusion from the south. This picture is consistent with geostrophic computations from the hydrographic section off Punta San Juan prior to the reversal of currents, and off Cabo Nazca during the northward flow. During the former the geostrophic longshore flow was poleward from the surface to below 200 m, with a maximum flow at 100 m. During the 'event' the poleward flow occurred only between the most offshore pair of stations—and equatorward flow was computed for the pair of stations just offshore and the pair over the slope.

Although surface countercurrents very close to the coast have been reported in eastern boundary current regions (Wooster and Reid, 1963), they are usually associated with the local allaying of upwelling (Gunther, 1938). The data presented here were during a period of upwelling, and the poleward flow appears to be more extensive than the 'surface countercurrent' of Gunther. The flow is more likely a manifestation of the Peru–Chile Undercurrent.

A COMPARISON

The description presented of coastal upwelling off Oregon has benefited from several years of study. A previous paper (Collins, *et al.* 1968) has shown the general features are repeated in observations in subsequent years. Caution is necessary in the comparison of the Oregon measurements with those made during only 2 weeks off the Peru Coast, and obviously during a 'quasibarotropic' event.

Nevertheless, both are regions of 'climatologically' predictable upwelling. The scalar mean speed and standard deviations are very similar. In both the surface Ekman layer appears to be shallow, i.e. less than 20 meters, and the flow below that layer appears to be quasi-geostrophic. Off Oregon, the strong density stratification (resulting from a strong halocline) leads to marked baroclinicity in the currents, while off Peru the minimal stratification (almost isohaline) allows the currents on the shelf to be barotropic.

There is unequivocal evidence of a poleward undercurrent on the continental shelf in both regions. Off Oregon it is covered by a surface current of about 40 meters in depth. At the time of the study off Peru, the undercurrent was only occasionally masked at the surface by a wind drift current.

The rather large standard deviations in the components of the current vectors measured off Peru are due to the 'event', the temporary reversal of the mean flow. Off Oregon the standard deviations reflect the large tidal and inertial currents present.

Off Peru there is no evidence for cross-stream flow and the sinking of upwelled water as it moved offshore, but the lack of marked indicators, i.e. the temperature inversion off Oregon, should not be construed to mean the flow pattern is not similar.

From observations of the distribution of physical and chemical properties it is apparent that the depth from which the water upwells is deeper off Oregon than Peru. The winds off Peru are steadier than off Oregon and the mean component in the direction favorable for upwelling is at least as strong. The relation between the intensity of stratification and the depth from which water upwells remains an intriguing question.

EPILOGUE

Cushing (1969), in substantial agreement with the remarks of Wooster quoted earlier, said in his review paper "Upwelling and Fish Production": "Apart from the theoretical work of Hidaka and Yoshida, the study of upwelling has been in the first stage of exploration and description. There is a second stage in oceanographic research where a survey should be mounted in sufficient detail to allow a more analytical approach to be sustained." Cushing sketched the type of survey needed, with emphasis on what should be done in biological and fishery oceanography. The Peru cruise, which provided the physical oceanography discussed above, was a first attempt at the type of survey where the biological, chemical and physical oceanographer worked together and in detail with the goal of understanding an upwelling area. A similar study off the Oregon Coast has just completed the field work phase. We should like to sketch what should be done from the aspect of physical oceanography in a future survey directed toward understanding the upwelling process.

The study should last for several months, but if it is to be done with sufficient detail the study will have to be spatially limited. The atmospheric systems appear to have significant variations with periods of the order of several days (cf. Mooers and Smith, 1967) and the velocities of the eastern boundary currents are of the order of 10 to 20 cm/sec. These numbers lead to a characteristic length of the order of 100 km if one wants to observe response of the water (or upwelling) to variations in the large scale atmospheric systems. Although the most active upwelling is very close to the coast, 20 km or so, the effects extend at least to the edge of the continental shelf. Since undercurrent may in some manner be 'tied' to the continental slope and may be a source of upwelled water, it would be advisable to include at least part of the slope region. So, an area some 100 km long by 50 km wide would be just adequate for an intensive mesoscale study off the west coast of North America, and a somewhat narrower region would suffice off Peru or Chile. It would be excellent if the study could be done as part of a larger scale survey of eastern boundary current, but the detailed observations in the mesoscale study should not be sacrificed in order to cover a greater area.

Classical hydrographic and STD costs will be necessary to define the field of density and the chemical properties. Although some ship time will be needed for servicing buoys, much of the ship time could be used for biological oceanography. For understanding the physical oceanography it will be important to consider the following:

1 Since the wind stress is the primary driving force for upwelling, the need for good meteorological data *over* the upwelling region is paramount. There should be at least one relatively large, stable buoy fully equipped in the meteorological instruments. Preferably there would be several at various distances from shore. Since upwelling influences the local meteorology and climate, this should be of considerable meteorological interest as well. There is, of course, the possibility of putting additional instrumentations, such as radiometers, on the buoys. The use of data buoy systems in coastal upwelling regions has been recently discussed by Wooster, Isaacs, and others (U. S. Coast Guard, 1969).

2 To define the currents and temperature (ideally density) field and their variations, and to provide the basic description of the water motion, 20 or more subsurface moored arrays of several recording current meters and thermographs each should be placed in the survey area. If the wind-drift layer is as shallow as the results presented above indicate, it would be easiest to measure and follow the very near surface flow with drogues. The time series of currents and temperature, coupled with the time series of the wind, would allow the upwelling region to be viewed as a dynamical physical system with meteorological and oceanographic inputs and outputs— and the various transfer or response functions to be obtained. This would be a major step toward obtaining a predictive scheme.

3 An airplane, equipped with an airborne radiation thermometer (ART), and perhaps a spectrophotometer for detecting the chlorophyll content of the surface water, would provide an overall detailed picture of the state of upwelling. This would be invaluable in following the development and changes of upwelling in real time and guiding surveys and experiments performed by the ships.

4 The possibility of directly measuring the vertical motion in an upwelling region is available. Two instruments recently developed are particularly promising—and may eliminate the need to 'guess' vertical velocities. A vertical current meter that is a rotating neutrally-buoyant float capable of sensing vertical motions of less than 1 m/hr (3×10^{-2} cm/sec) has been built and used at sea (Webb and Worthington, 1968; Voorhis, 1968). It will be used in an upwelling region this next year for the first time by Voorhis, Webb and Ryther of the Woods Hole Oceanographic Institution. It works best as a vertical current meter in minimally stratified water. For use in more highly stratified water the Autoprobe (Winn *et al.*, 1968), a free platform capable of adjusting its displacement to operate in an isothermal (or any other suitable measured parameter) mode could be tracked acoustically. Tracking the autoprobe in an upwelling region might allow the trajectory of a particular water parcel in the upwelling region to be determined.

5 While our knowledge of the upwelling process at present appears to be data limited, the importance of theory should not be neglected. Until recently, oceanographers have not developed numerical models to the extent meteorologists have; meteorologists have had appreciable success in numerical weather prediction. Dr. J. J. O'Brien of Florida State University has recently begun a numerical model of upwelling on a coastal shelf and one can eventually hope to have predictive models of the upwelling region.

All of the elements of the physical oceanography program above are now being performed, but they are for the most part being done separately, and in different parts of the ocean, and not necessarily focused on upwelling. The greatest task would be that of assembling and coordinating the various investigations to focus on a single upwelling region in concert with the biological, the chemical and the fisheries oceanographers. It is gratifying to see that studies of coastal upwelling are given priority in the U. S. discussion of the International Decade of Ocean Exploration (National Academy of Sciences, 1969).

Acknowledgements

We wish to acknowledge our debt to R. Dale Pillsbury, Oregon State University, for his assistance and advice, and for his effort in preparing, installing, and recovering so many instruments so successfully. Much guidance and assistance was provided by Professors June G. Pattullo, Oregon State University, and Richard C. Dugdale, University of Washington, in the studies off Oregon and off Peru, respectively. This paper is a partial synthesis of results of research supported by the Office of Naval Research contract Nonr 1286 (10), the U. S. Bureau of Commercial Fisheries contract 14-17-0002-333, and the National Science Foundation grants GA331 und GA1435, to Oregon State University, and GB8648 to the University of Washington.

References

ARMSTRONG, F. A. J., STEARNS, C. R. and STRICKLAND, J. D. H. (1967). The Measurement of Upwelling and Subsequent Biological Processes by Means of the Technicon Autoanalyzer and Associated Equipment. *Deep-Sea Res.* 14, 381–398.

ARTHUR, R. S. (1965). On the calculation of vertical motion in eastern boundary currents from determinations of horizontal motion. *J. Geophys. Res.* 70, 2799–2804.

COLLINS, C. A., MOOERS, C. N. K., STEVENSON, M. R., SMITH, R. L., and PATTULLO, J. G. (1968). Direct Current Measurements in the Frontal Zone of a Coastal Upwelling Region. *J. Oceanogr. Soc. Japan* 24, 295–306.

CUSHING, D. H. (1969). Upwelling and Fish Production. *FAO Fisheries Technical Paper* 84. Rome: Food and Agricultural Organization of the United Nations.

EBER, L. E., SAUR, J. F. T., and SETTE, O. E. (1968). Monthly Mean Charts Sea Surface Temperature North Pacific Ocean 1949–1962. Washington, D. C.: U. S. Bureau of Commercial Fisheries (Circular 258).

EKMAN, V. W. (1905). On the Influence of the Earth's Rotation on Ocean Currents. *Ark. Mat. Astr. Fys.* 2, 11, 1–52.

GUNTHER, E. R. (1936). A Report on Oceanographical Investigations in the Peru Coastal Current. *Discovery Rep.* 13, 107–276.

McEWEN, G. F. (1912). The Distribution of Oceanic Temperatures Along the West Coast of North America Deduced from Ekman's Theory of the Upwelling of Cold Water from the Adjacent Ocean Depths. *Int. Revue Ges. Hydro. Biol. Hydrogr.* 5, 243–286.

MOOERS, C. N. K. (1969). The Interaction of an Internal Tide with the Frontal Zone of a Coastal Upwelling Region. Ph. D. thesis, Oregon State University.

MOOERS, C. N. K. and SMITH, R. L. (1967). Continental Shelf Waves off Oregon. *J. Geophys. Res.* 73, 549–557.

National Academy of Sciences, 1969. *An Oceanic Quest.* Washington, D. C.

PAK, H., BEARDSLEY, G. F., and SMITH, R. L. (1970). An Optical and Hydrographic Study of a Temperature Inversion off Oregon During Upwelling. *J. Geophys. Res.* 75, 629-636.

PILLSBURY, R. D., SMITH, R. L., and TIPPER, R. C. (1969). A reliable low-cost mooring system for oceanographic instrumentation. *Limnol. Oceanogr.*, 14, 307–311.

RYTHER, J. H. (1969). Photosynthesis and Fish Production in the Sea. *Science* 166, 72–76.

SMITH, R. L. (1968). Upwelling. In *Oceanogr. Mar. Biol. Ann. Rev.* **6**, 11–47 (ed. Barnes, H) London: George Allen and Unwin Ltd.

SVERDRUP, H. U. (1938). On the Process of Upwelling. *J. Mar. Res.*, **1**, 155–164.

U. S. Coast Guard, 1969. Proceedings of the First USCG National Data Buoy Systems Scientific Advisory Meeting.

VOORHIS, A. D. (1968). Measurements of Vertical Motion and Partition of Energy in the New England Slope Water. *Deep-Sea Res.*, **15**, 599–608.

WEBB, D. C. and WORTHINGTON, L. V. (1968). Measurements of Vertical Water Movement in the Cayman Basin. *Deep-Sea Res.*, **15**, 609–612.

WINN, A. L., WEBB, D. C. and BURT, K. (1968). Autoprobe, a Platform for Midwater Observation. *Marine Science Instrumentation*, **4**, 347–353.

WOOSTER, W. S. and GILMARTIN, M. (1961). The Peru–Chile Undercurrent. *J. Mar. Res.*, **19**, 97–122.

WOOSTER, W. S. and Reid, J. L. (1963). Eastern Boundary Currents. In *The Sea*, Vol. **2**, 253–280 (ed. Hill, M. N.), New York: Interscience.

WYRTKI, K. (1963). The Horizontal and Vertical Field of Motion in the Peru Current. *Bull Scripps Instr. Oceanogr.*, **8**, 313–346.

YOSHIDA, K. (1967). Circulation in the Eastern Tropical Oceans with Special References to Upwelling and Undercurrents. *Jap. J. Geophys.*, **4**, 2, 1–75.

Nitrogen fixation in the sea

W. D. P. STEWART

Department of Biological Sciences,
University of Dundee, Scotland

In 1886–1888 the German workers Hellriegel and Wilfarth published evidence from studies on legumes which showed conclusively that N_2 from the atmosphere could be converted into combined form by biological agents. This process of nitrogen fixation is now known to be characteristic of a variety of terrestrial legumes, nodulated non-legumes, blue-green algae and bacteria and it is considered that over 90 per cent of the total nitrogen added to the surface of the Earth is biological in origin and that nodulated leguminous plants are the major contributors (see Stewart, 1966). Despite the facts that man has known of nitrogen fixation for such a long period and that land covers only about one fifth of the Earth's surface, it is only within the last decade that biological nitrogen fixation in the sea has been established with certainty. The information which I will present in this paper, although current, is thus fragmentary.

The lack of data in the past on marine nitrogen fixation is perhaps not unexpected. First, the techniques and equipment for measuring *in situ* nitrogen fixation have only been developed fairly recently. Second, the sea has been considered as an unlikely habitat for biological nitrogen fixation because nodulated angiosperms which contribute the bulk of combined nitrogen on land are entirely absent and third, the potential nitrogen-fixing plants which do occur, the blue-green algae and the bacteria, are of limited, distribution. In this paper I should like (1) to consider the methods of measuring nitrogen fixation in the sea, (2) to list the known nitrogen-fixing organisms and their quantitative significance, (3) to discuss some of the factors which may affect specifically nitrogen fixation in marine habitats and (4) to consider briefly the production of extracellular nitrogen by marine nitrogen-fixing organisms.

METHODS OF MEASURING NITROGEN FIXATION

Early indications that nitrogen-fixing microorganisms occur in marine environments were based on the isolation from the sea of species belonging to the known nitrogen-fixing genera *Azotobacter* and *Clostridium* but few of these species were critically tested for nitrogen fixation. Waksman, Hotchkiss and Carey (1933) summed up the position of the 1930's when they wrote "The function of the nitrogen-fixing bacteria in increasing the supply of nitrogen in the sea is difficult to determine, the results obtained under laboratory conditions not lending themselves directly to a direct application to natural marine processes". Since then techniques of measuring *in situ* nitrogen fixation have become available and the detection of nitrogen fixation in the sea is no longer a methodological nightmare. The most satisfactory method is to use the heavy isotope of nitrogen, ^{15}N, as tracer (Dugdale, Menzel and Ryther, 1961; Dugdale, Goering and Ryther, 1964; Stewart, 1965, 1967) but more recently it has been shown that the acetylene reduction technique may have particular application in aquatic ecosystems (Stewart, Fitzgerald and Burris, 1967, 1968).

The use of ^{15}N in studies on biological nitrogen fixation was initiated in laboratory studies by Burris, Eppling, Wahlin and Wilson (1943). The method depends on exposing the test sample to a gas phase containing $^{15}N_2$, and after a suitable exposure period, which may range from a matter of hours to a matter of days, the test material is then assayed for ^{15}N incorporation by mass spectrometry. Nitrogen-fixing plants do not discriminate between the various isotopes of nitrogen and it is generally accepted (Wilson, 1958) that a ^{15}N-enrichment of 0.015 atom per cent excess ^{15}N over an unexposed biological control constitutes satisfactory evidence of nitrogen fixation. Much lower ^{15}N enrichments (0.003–0.004 atom per cent excess ^{15}N) can be considered significant however if the replication between samples is good and if the methodology is sound. Burris and Wilson (1957) and Bremner, Cheng and Edwards (1965) have dealt in detail with the methodology of the ^{15}N method.

The ^{15}N technique used by Dugdale and co-workers in fresh-water ecosystems (Neess *et al.*, 1962) has been used by Dugdale's group in extensive simulated *in situ* experiments in certain oceanic waters where species of the blue-green alga *Trichodesmium* were abundant. The technique can be summarised as follows:

1 The test material is added to natural sea water in a 1.0 l boiling flask which is fitted with a gas flushing unit.

2 Air is removed from the water with a premixed N_2-free gas phase at a pressure of 0.8 atm. Neess *et al.* (1962) used a gas mix of O_2/He of 20/80 but Fogg and Horne (1967) recommended the addition of CO_2 and used a flushing mixture of argon/O_2/CO_2 of 75/25/0.04.

3 Nitrogen gas enriched with ^{15}N is added to the flask (0.3 atm.) and the gas is equilibrated with the aqueous phase by shaking.

4 The test sample is then incubated under the desired experimental conditions for a known period, generally 6–24 h.

5 The nitrogen in the test sample is converted to nitrogen gas and its ^{15}N content is determined by mass spectrometry.

6 The ^{15}N-labelling of the test sample is compared with the ^{15}N-labelling of a sample which has not been exposed to $^{15}N_2$ to determine whether uptake of the isotope has occurred and the nitrogen fixed is calculated as follows:

$$\text{Nitrogen fixed} = \frac{\text{Atom per cent excess } ^{15}N \text{ in test sample}}{\text{Atom per cent excess } ^{15}N \text{ in gas phase}}$$

$$\times \text{ total nitrogen per test sample}$$

Stewart (1967) introduced a modification of this method in his studies on nitrogen fixation by epilithic blue-green algae but this differs only in detail from Dudgale's technique and particularly in the simplicity with which the $^{15}N_2$ can be stored and transported.

In general the ^{15}N method is the most reliable method of measuring nitrogen fixation but it has disadvantages in that it is expensive, requires complicated equipment including a mass spectrometer and it requires considerable operator skill. The use of the acetylene reduction technique in studies on aquatic nitrogen fixation has largely overcome many of these difficulties. This technique is based on the discovery by Schöllhorn and Burris, (1966, 1967) and by Dilworth (1966) that the nitrogen-fixing enzyme complex nitrogenase will reduce a variety of substrates in addition to N_2. These are listed in Table I. (See Hardy and Burns, 1968.) The majority of

15*

Table I Substrates reduced by the nitrogen-fixing enzyme complex, nitrogenase

Nitrogen gas	N_2	$\rightarrow$	$2\,NH_3$
Nitrous oxide	N_2O	$\rightarrow$	$N_2 + H_2O$
Azide	N_3^-	$\rightarrow$	$N_2 + NH_3$
Acetylene	C_2H_2	$\rightarrow$	C_2H_4
Cyanide	HCN	$\rightarrow$	$CH_4 + NH_3 + CH_3NH_2$*
Isocyanide	CH_3NC	$\rightarrow$	$CH_4 + C_2H_4$* $+ C_2H_6$* $+ CH_3NH_2$*

* Denotes traces only formed.

these substrates inhibit whole cell metabolism and with the exception of acetylene are thus unsuitable for quantitative studies on intact organisms. The acetylene reduction technique as used by Stewart, Fitzgerald and Burris (1967) is summarised below, together with some comments which are relevant in the light of our experience with the method:

1 A 1.0 ml sample of the alga to be tested is added to a 5–7 ml capacity serum bottle fitted with a serum stopper.

2 Air is removed from the bottle by flushing with a pre-mixed gas phase of $A/O_2/CO_2$ (78/22/0.04).

3 Ten per cent purified acetylene is added and the bottles are incubated *in situ* for a 30 min. period at the end of which the reaction is terminated by the addition of 0.2 ml of 50 per cent trichloroacetic acid.

4 A sample of the gas phase is analysed for ethylene production by gas chromatography, using a hydrogen-flame ionisation detector, and the ethylene produced gives a direct measure of the nitrogenase activity in the test sample.

Several modifications of this basic technique may find application in marine studies. First, it will generally be necessary to concentrate the phytoplankton prior to exposure to acetylene, and with the macroalgae, larger bottles will obviously be required. The volume of the container used is immaterial. Second, we have found with freshwater plankton samples that it is not necessary to remove the air if 20% acetylene is added to the gas phase. Then the rate of acetylene reduction is the same as when 10% acetylene is added in the absence of N_2. The effect of increasing the gas pressure by adding 20% acetylene can be overcome by pricking the serum stopper with a hypodermic needle. This modification is particularly useful because it is possible to expose a test sample to acetylene without removing the *in situ* gas phase. Third, we have found in field studies with *Calothrix* as well as with some terrestrial soils that the addition of acid, including trichloro-acetic acid, results in the spontaneous production of ethylene when acetylene is present. In these instances we terminate the reaction with 5% sodium azide, or if total nitrogen data are required we withdraw a 1.0 ml sample of the gas phase into a syringe at the end of the experiment, seal this with a rubber stopper and subsequently analyse the gas phase back in the laboratory. Hardy *et al.* (1968) have also used a system of syringes in which they in-cubate the test samples and store the gas phases.

The acetylene reduction technique is particularly suitable for studies on aquatic ecosystems because in addition to it being rapid, extremely sensitive and simple, the substrate (acetylene) is highly soluble in water while the product (ethylene) is very insoluble in water. The outstanding problem in relation to its use is that it is not yet known with certainty whether the rate of acetylene reduction: nitrogen reduction is constant under all experimental conditions. Experimental data to date with cell-free extracts of *Azotobacter* and *Clostridium* (Schöllhorn and Burris, 1967; Klucas, 1968; Hardy *et al.*, 1968) as well as with blue-green algae (Stewart, Fitzgerald and Burris, 1968) indicate that the ratio is near to 3:1. If this ratio proves to be constant under all experimental conditions it will be possible to measure acetylene reduction, apply a constant and arrive at a measure of the rate of nitrogen fixation.

THE MARINE ORGANISMS WHICH FIX NITROGEN

Known nitrogen-fixing angiosperms, as noted earlier, are absent from the sea and there is nothing to suggest that marine grasses such as *Zostera* fix nitrogen. From analogy with terrestrial ecosystems it seems that if nitrogen fixation is to occur in the sea it will be due to the activity of algae or of free-living bacteria.

1) Algae

Early reports in the literature suggest that a variety of algae such as species of *Enteromorpha*, and various coccoid green algae, including *Chlorella*, fix nitrogen, but Fogg (1956) in reviewing the situation concluded that there was no good evidence of nitrogen fixation by algae other than the Myxo-phyceae. He did stress, however, that few species had been tested. Using the acetylene reduction technique we have tested a variety of temperate

marine intertidal algae for nitrogenase activity. The algae were taken directly
from the seashore and exposed to acetylene in the laboratory within 45 min.
of collection. Samples were also assayed for oxygen production using the
Winkler technique and only acetylene reduction data obtained for metabolically active samples were considered. None of the species of Chlorophyta, Rhodophyta or Phaeophyta tested showed nitrogenase activity and
these are listed in Table II. Samples of epilithic *Calothrix* species sampled
from the same general area at the same time as most of the macro-algae
(October, 1969) reduced acetylene vigorously (Table III). On one occasion
nitrogenase activity was detected in a sample of *Chaetomorpha* but subsequent microscopic examination showed that *Calothrix* was present as an
epiphyte on the alga and when the organisms were separated and then
retested it was found that *Calothrix* and not *Chaetomorpha* was responsible
for the observed nitrogenase activity (Table IV). These negative data for
the larger algae infer indirectly that epiphytic nitrogen-fixing bacteria such
as *Azotobacter* are unimportant as contributors of combined nitrogen in
the habitat from which we obtained our algal samples.

Table II   Marine macro-algae which are unable to reduce acetylene to ethylene

Ascophyllum nodosum	*Halidrys siliquosa*
Chaetomorpha sp.	*Laminaria digitata*
Chondrus crispus	*Laminaria saccharina*
Cladophora rupestris	*Laurencia pinnatifida*
Corallina officinalis	*Pelvetia canaliculata*
Ectocarpus siliculosus	*Pilayella littoralis*
Enteromorpha compressa	*Polyides rotundus*
Enteromorpha linza	*Polysiphonia urceolata*
Fucus serratus	*Porphyra leucosticta*
Fucus spiralis	*Porphyra linearis*
Fucus vesiculosus	*Prasiola stipitata*
Furcellaria fastigiata	*Rhodymenia palmata*
Gigartina stellata	*Ulothrix sp.*
	Ulva lactuca

The taxonomy used in this table is that of Parke and Dixon (1968). Triplicate samples
of each alga were tested. The methodology was based on that of Stewart *et al.* (1967)
although containers with a volume of up to 250 c.c. were used with some of the larger
algae. In the case of the *Laminaria* species, samples of the thallus were used rather
than intact plants.

Table III Acetylene reduction by epilithic *Calothrix* samples

Sample	n moles C_2H_4/mg. protein/min
1	0.16
2	0.05
3	0.18
4	0.11
5	0.08
6	0.13

The methods used were as described by Stewart *et al.*, 1967 except that the experiment was not terminated by the addition of acid but by withdrawing a gas sample into a syringe at the end of the 30 min. incubation period. A control series showed that in the absence of acetylene there was no ethylene production.

Table IV Tests for acetylene reduction by a *Chaetomorpha* species bearing an epiphytic *Calothrix* species

Sample	No.	n mol. C_2H_4/mg. protein/min	
Chaetomorpha + *Calothrix*	1	0.01	
	2	0.01	0.01
	3	0.01	
Chaetomorpha alone	1	0.00	
	2	0.00	0.00
	3	0.00	
Calothrix alone	1	0.22	
	2	0.25	0.26
	3	0.32	

Methods used were as in Stewart *et al.*, 1967.

The lack of positive data for the macro-algae and the positive data for blue-green algae led us to investigate a variety of marine blue-green algae for nitrogenase activity. The species for which positive data have been obtained are listed in Table V. Species 1–14 have been tested in my laboratory. *Calothrix scopulorum* and *Nostoc entophytum* were available in pure culture and it has been shown unequivocally that they fix nitrogen. Species 3–9 were available in unialgal culture and grew vigorously in combined nitrogen-free medium. Species 10–14 were collected as natural populations and perhaps other nitrogen-fixing contaminants contributed to the observed nitrogen

Table V Marine blue-green algae for which there is evidence of nitrogen fixation

No.	Alga	Sample	Origin	Method	Mean n moles C_2H_4/ mg.prot./min.	Mean atom% excess ^{15}N	Reference
1.	*Calothrix scopulorum*	pure	supralittoral, west coast of Scotland	^{15}N	—	—	Stewart (1964)
2.	*Nostoc entophytum*	pure	supralittoral, west coast of Scotland	^{15}N	—	—	Stewart (1964)
3.	*Anabaena torulosa*	unialgal	salt marsh on English east coast	^{15}N	—	1.252	—
4.	*Anabaena variabilis*	unialgal	salt marsh on English east coast	^{15}N	—	0.928	—
5.	*Calothrix aeruginea*	unialgal	supralittoral, west coast of Scotland	^{15}N	—	0.323	—
6.	*Microchaete* sp.	unialgal	Butcher Collection, Westfield College, London	C_2H_2	0.52	—	—
7.	*Nodularia harveyana*	unialgal	salt marsh on English east coast	^{15}N	—	1.285	—
8.	*Nodularia spumigena*	unialgal	salt marsh on English east coast	^{15}N	—	0.927	—
9.	*Nostoc linckia*	unialgal	supralittoral, west coast of Scotland	^{15}N	—	1.906	—
10.	*Nostoc* sp.	field	supralittoral, west coast of Scotland	^{15}N	—	0.117	—
11.	*Calothrix confervoides*	field	upper littoral, west coast of Scotland	^{15}N	—	0.056	—

12.	*Nostoc* sp.	field	supralittoral, west coast of Scotland	^{15}N	1.342	—	—
13.	*Rivularia atra*	field	supralittoral rock face, west coast of Scotland	^{15}N	0.083	—	—
14.	*Rivularia biasolettiana*	field	supralittoral rock face, coast of Wales	^{15}N	0.249	—	—
15.	*Calothrix crustacea*	pure	California coastal water	total N	—	—	Allen (1963)
16.	*Nostoc* sp. (*Anabaena* sp.)	pure	California coastal water	total N	—	—	Allen (1963)
17.	*Trichodesmium erythraeum*	field	tropical oceanic	^{15}N	—	—	Dugdale et al. (1961, 1964)
18.	*Trichodesmium thiebautii*	field	tropical oceanic	^{15}N	—	—	Dugdale et al. (1961, 1964)

The exact enrichments and experimental details for species 1, 2 and 15–18 are given in the original publications. The exposure periods to ^{15}N$_2$ and the atom per cent ^{15}N in the gas phases to which the other algae were exposed were: species 3, 4, 7, 8 (12 h. to 90 atom per cent ^{15}N); species 5 (6 h. to 90 atom per cent ^{15}N); species 9 (12 h. to 60 atom per cent ^{15}N); species 10–13 (6 h. to 95 atom per cent ^{15}N); species 14 (12 h. to 60 atom per cent ^{15}N). The *Microchaete* species was analysed for acetylene reduction by the method of Stewart *et al.*, 1967. Each value given is the mean of triplicate determinations.

fixation. However this seems unlikely because samples of non-heterocystous algae such as *Lyngbya, Oscillatoria, Phormidium* and unicellular forms sampled from adjacent areas showed no evidence of nitrogenase activity. Species 15–18 are organisms which have been tested by other workers. Allen (1963) reported on nitrogen fixation by two strains of the marine species *Calothrix crustacea* and by a marine strain of *Nostoc* (or *Anabaena*) which she isolated on synthetic sea water media.

The *Calothrix* strain fixed 0.23–0.24 mg N/100 ml medium/day over a 60-day period and the corresponding value for *Nostoc* was 0.09 mg N. Dugdale *et al.*, (1961, 1964) have reported on nitrogen fixation by *Trichodesmium* and Ramamurthy and Kirshnamurthy (1968) claim that they have isolated pure cultures of this alga which grow better on N_2 than on nitrate, or ammonium-nitrogen. The success of Ramamurthy and Krishnamurthy is remarkable because many other workers have failed to grow the alga, let alone obtain bacteria-free cultures. Dugdale's work on this alga will be considered in detail later.

In summation, there are now data which indicate that 18 species or strains of marine blue-green algae fix nitrogen. All are members of the orders Nostocales or Stigonematales and with the exception of *Trichodesmium*, all possess heterocysts. The direct correlation between the presence of heterocysts and a potential nitrogen-fixing capacity is particularly interesting from an ecological point of view because simply by determining the relative abundance of heterocysts one may arrive at a measure of the nitrogen-fixing potential of a particular aerobic flora.

It is fair to say that although an appreciable number of temperate marine blue-green algae have been tested for nitrogen fixation a large number still remain to be tested. This includes virtually all tropical species, except *Trichodesmium*, but including in particular the *Nostoc* colonies of coral reefs, the gelatinous blue-green algae of intertidal regions, the blue-green algae of the Florida everglades and the *Nostoc* species recorded by Bernard and Leçal (1960) from various depths of the Atlantic, Mediterranean and Pacific Oceans, as well as the *Anabaena* species which sometimes occur in the plankton of the North Sea (R. Johnston, personal communication).

Before leaving the blue-green algae one must consider the possibility of nitrogen fixation by marine lichens. The results for 9 species from marine or coastal regions (Table VI) show that nitrogen fixation occurs only in those species which contain blue-green algae as phycobionts. The ability of *Lichina pygmaea* to fix nitrogen is interesting because this lichen may extend down to mid-tide level of the seashore. On the east coast of Scotland *L. confinis* and *L. pygmaea* both fix nitrogen in winter at temperatures near to 0°C and do so at higher rates than do free living epilithic *Calothrix* species under similar conditions. Lichens may thus form an important source of combined nitrogen in some habitats in winter.

Table VI   Tests for uptake of $^{15}N_2$ by coastal and marine lichens

Lichen	Phycobiont	Mean Atom % excess ^{15}N
Parmelia conspersa	*Trebouxia*	+0.001
Parmelia fuliginosa	*Trebouxia*	−0.001
Parmelia saxatilis	*Trebouxia*	+0.003
Physcia adscendens	*Trebouxia*	+0.003
Ochrolechia parella	*Trebouxia*	0.000
Xanthoria parietina	*Trebouxia*	0.000
Collema auriculatum	*Nostoc*	+0.480
Lichina confinis	*Calothrix*	+0.454
Lichina pygmaea	*Calothrix*	+0.117

The lichens were collected on April 20th, 1965 from the west coast of Wales and exposed to $^{15}N_2$ enriched with approximately 30 atom per cent excess ^{15}N for 5 days, at 20°C. and in alternating periods of 12 hr light (2500 lux) and 12 hours dark. The initial gas phase contained N_2 (20%), O_2 (20%), CO_2 (0.04%) and the balance was argon. Each result given is the mean of triplicate determinations.

2) Photosynthetic bacteria

The role of photosynthetic nitrogen-fixing bacteria in the sea is unknown because no marine species have apparently been tested for nitrogenase activity. However fresh water strains of *Chlorobium*, *Chromatium*, *Pelodictyon*, *Rhodomicrobium*, *Rhodopseudomonas* and *Rhodospirillum* all fix nitrogen (see Stewart, 1966, 1968) so that there is no reason to believe that marine photosynthetic bacteria do not possess a similar capacity. They are certainly of widespread distribution in marine habitats, particularly in salt marshes, estuarine muds, sediments, in temporary tidal pools and among decaying larger algae (see Zobell, 1946; Kondrat'eva, 1965). It is important in relation to nitrogen fixation to remember that although most photosynthetic bacteria grow under microaerophilic conditions they fix N_2 only under anaerobic conditions. This emphasises the importance of carrying out *in situ* tests rather than trying to determine their importance on the basis of laboratory counts. In terrestrial ecosystems photosynthetic bacteria may be more important than blue-green algae in some soils (Kobayashi, Takahashi and Kawaguchi, 1967) and Kobayashi (1968) has estimated that there these organisms may contribute between 3.0–30.0 mg N/m²/ann.

3) Heterotrophic bacteria

As on land little is known about the ecological significance of heterotrophic nitrogen-fixing bacteria but because of the shortage of available energy sources in marine environments it is unlikely that they are of very great importance except in certain restricted environments. The first report of their occurrence in marine habitats was that of Benecke and Keutner (1903) who in studies on the Baltic isolated *Azotobacter* from among the marine plankton and *Clostridium* from the deeper waters and bottom sediments. This was confirmed by various other early workers (see Waksman, Hotchkiss and Carey, 1933 for references). The latter workers in summarising the earlier findings, and their own, concluded that both *Azotobacter* and *Clostridium* were widely distributed in the marine environments and that there were true marine forms, not simply fresh-water types which had been washed into the sea from the land. In addition to *Azotobacter* and *Clostridium* a variety of other known nitrogen-fixing organisms exist in the sea but few have been critically evaluated for nitrogen fixation in the laboratory, let alone for their ability to fix nitrogen *in situ*.

   The abundance and distribution of *Azotobacter* in the Black Sea has been studied extensively by Pshenin (1963). In samples from over 90 stations, he isolated 60 *Azotobacter* strains belonging to the species *A. chroococcum*, *A. vinelandii*, and *A. nigricans*. The strains which he tested for nitrogen fixation were capable of doing so, the average efficiency of the cultures being 6.5 mg nitrogen fixed/1.0 gm glucose consumed. *Azotobacter* was distributed throughout the water column, even in the H_2S-zone. He suggested that moribund phytoplankton and zooplankton excreta were important sources of carbon, as were the bottom sediments and various macro-algae such as *Phyllophora*, *Cystoseira*, *Ulva* and *Enteromorpha*. The occurrence of *Azotobacter* as an epiphyte on macro-algae was also noted by early workers (see Waksman, Hotchkiss and Carey, 1933; Zobell, 1946). Whether the *Azotobacter* species fixed nitrogen *in situ* was not investigated in any of these studies but it is notable that we found no evidence of nitrogenase activity in our studies on the larger marine algae in Scotland indicating that there nitrogen fixation by heterotrophic bacteria, if it occurred, was unimportant. Our findings thus do not support the view of Reinke (1903) who concluded that the association of *Azotobacter* with algae may be a symbiotic one with the algae providing carbohydrates for the bacteria, and the bacteria supplying combined nitrogen for the algae.

The nitrogen-fixing capacity of strains of *Clostridium*, particularly *Clostridium pasteurianum*, is well-known but not all strains of *Clostridium* fix nitrogen, and there are no quantitative data on their nitrogen-fixing activity in the sea. Pshenin (1963) concludes that they are important components of the sediments of the Black Sea from which he isolated about 30 strains. The nitrogen-fixing efficiency of the marine isolates are low but nevertheless these marine strains, if they fix nitrogen *in situ*, must contribute to the nitrogen content of the ecosystem.

The finding of Hill and Postgate (1969) that a strain of *Pseudomonas* fixes nitrogen is of particular interest because this genus is of widespread and abundant occurrence in the sea. In addition to *Pseudomonas* it has also been confirmed that some *Desulfovibrio* strains fix nitrogen. Sisler and Zobell (1951) were the first to demonstrate nitrogen fixation by *Desulfovibrio*. They tested their marine isolate using a nitrogen/argon ratio method. This depends on exposing the test sample to a gas phase which includes N_2 and argon. Argon which is a biologically inert gas, is used as an internal standard. A decrease in the nitrogen/argon ratio of the gas phase, when measured by mass spectrometry, is indicative of N_2 assimilation by the test sample. This method is less sensitive than either the ^{15}N method or the acetylene reduction technique and has not been widely applied. Although Sisler and Zobell's strain of *Desulfovibrio* fixed nitrogen this is not characteristic of many strains and only *in situ* determinations can tell of their actual significance in marine sediments where they are widely distributed.

Species of the genus *Mycobacterium* were first reported to fix nitrogen by Fedorov and Kalininskaya (1961 *a*, *b*, *c*) and three nitrogen-fixing species *M. flavum* 301, *M. roseo-album* 368 and *Mycobacterium* sp. 571 are known. Nitrogen fixation by the first species has been studied in some detail by Biggins and Postgate (1969). The ability of *Mycobacterium* species to fix N_2 is of particular interest because this genus is common and sometimes abundant in the Indian Ocean, Pacific Ocean, Arctic Ocean, Greenland Sea, and the Black Sea (Kriss *et al.*, 1967). However no data are available on whether these marine strains fix nitrogen, or whether they are active in the sea.

Marine *Spirillum* species appear to be more important than either *Azotobacter* or *Clostridium* in the Black Sea (Pshenin, 1963) and the efficiency of these isolates in fixing nitrogen is 4.2–16.1 mg N fixed/gm carbohydrate consumed (mean 8.7) under Pshenin's conditions. This surpasses the efficiency of the *Azotobacter* strains tested under similar conditions. Kriss *et al.* (1967) find spirillae to be of widespread occurrence in many seas.

4) Yeasts

The possibility of nitrogen fixation by marine yeasts has been considered by Allen (1963). In routine screening tests of sea water taken from the coast off Northern California, she isolated pink yeasts of the genus *Rhodotorula* which grew on "strictly nitrogen-free liquid media" and she considered that there was strong presumptive evidence that these yeasts fixed nitrogen, but as far as this author is aware, no confirmatory evidence for this has been published. Pshenin (1963) reported on nitrogen fixation by strains of *Rhodotorula* and *Torulopsis*, but the reported efficiencies (mg N fixed/gm glucose consumed) are 0.25 and 1.0 respectively. These efficiencies are very low and such findings should be confirmed in other laboratories. This is particularly important in view of the recent paper by Millbank (1969) who failed, using critical methods, to confirm the earlier reports of nitrogen fixation by terrestrial strains of *Rhodotorula* and *Pullularia*.

IN SITU MEASUREMENTS OF MARINE NITROGEN FIXATION

To date *in situ* measurements of nitrogen fixation in the sea appear to be confined to studies on *Trichodesmium* and on *Calothrix*, although work on sand dune slack blue-green algae has also been carried out. Fortunately *Trichodesmium* is a tropical oceanic planktonic species and *Calothrix* is an epilithic temperate supralittoral and intertidal form so that one set of data complements the other.

Trichodesmium species are characteristically reddish or brownish in color due to the excessive development of the red pigment c-phycoerythrin. Taxonomically the genus is placed in the family Oscillatoriaceae and in fact Geitler (1932) considers it to be a species of *Oscillatoria*. Another tropical marine planktonic blue alga *Katagnymene* resembles *Trichodesmium* but the trichomes do not occur in bundles. There is some controversy as to whether this is really a distinct genus, or whether it is simply a form of *Trichodesmium*. *Trichodesmium* occurs in the plankton in thermally stratified nutrient-poor waters and although it is found particularly in tropical waters it has been noted as far north as the south coast of Ireland in summer (Farran, 1932) presumably having been carried there by the Gulf Stream. At the other extreme Wood (1965) reports it off the South Island of New Zealand. The distribution of the alga has been dealt with in several publications (Delsman, 1939; Brongersma–Sanders, 1948, 1957).

Dugdale, Menzel and Ryther (1961) first reported on nitrogen fixation by natural populations of *Trichodesmium*. In studies off the south coast of Bermuda, they concentrated *Trichodesmium* samples by towing and inoculated the alga in the laboratory into flasks which contained surface sea water. The algae were then exposed to $^{15}N_2$ and incubated under simulated *in situ* conditions in a water-cooled bath fitted with neutral density filters. After a 5 h exposure period to ^{15}N it was found that nitrogen fixation had occurred in 3 out of the 4 test flasks and that there was a direct correlation between the decrease in light intensity and decrease in the rate of nitrogen fixation, with no detectable nitrogen fixation occurring in the dark. In subsequent studies in the Atlantic Ocean, Indian Ocean and the Arabian seas Dugdale, Goering and Ryther (1964) obtained rates of *in situ* nitrogen fixation of up to 2 μg N/1/hr, with the average spring and summer rates being near to 0.15 and 0.05 μg N/1/hr respectively. These high rates of nitrogen fixation were on occasion up to 10 times greater than the rate of removal of ammonia from the sea by phytoplankton so that a considerable input of *new* nitrogen into the ecosystem must result. Dugdale and Goering (1967) conclude that nitrogen fixation can support growth of *Trichodesmium* at rates comparable to that obtained on nitrate-nitrogen.

As Dugdale *et al.* (1964) point out the observed nitrogen fixation is probably due to the *Trichodesmium* but it is possible that associated microorganisms may be responsible. From the ecological point of view this is immaterial—the important thing is that nitrogen fixation occurs and that this may be of considerable significance, particularly when blooms of the alga in the Pacific may on occasion cover areas of about 20,000 square miles (Wood, 1965). The occurrence of such large blooms in oceanic waters is difficult to reconcile with the view of Feldman (1932) that *Trichodesmium* is in fact a bottom-living alga which occasionally comes to the surface. The reddish coloration is however characteristic of certain bottom-living blue-green algae (Pintner and Provasoli, 1958) and it would be interesting to know whether *Trichodesmium* like many other marine Oscillatoriaceae (Van Baalen, 1962) has a vitamin B_{12} requirement.

Stewart (1965, 1967) studied *in situ* nitrogen fixation by epilithic algae on the Scottish coast and by sand-dune slack algae on the east coast of England using ^{15}N as tracer. The algae which dominated the supralittoral fringe of the rocky shore were principally *Calothrix* species. In preliminary experiments appreciable fixation was detected in February even although the temperatures at that time of the year were generally about 0–10°C. Fixation was considerably higher in the light than in the dark indicating that it was largely algal in origin. Subsequent studies on seasonal variation (Table VII) showed that nitrogen fixation per unit area was highest in spring, lower in autumn, and there was little fixation in summer and in winter. It can be seen that during the period of rapid colonization in August and September nitrogen fixation was responsible for between 27–37 per cent of the total increase in nitrogen while in spring (March–April) the percentage was approximately 30 per cent. At other times of the year the percentage increase in total nitrogen due to nitrogen fixation was much less. The total nitrogen fixed in this habitat was approximately 2.5 gm/m²/annum which is about one-tenth of that fixed by a good leguminous crop in a year, or put in another way, the total nitrogen fixed per annum was equivalent to 41% of the mean total nitrogen present per annum in this particular habitat. Thus the contribution of *Calothrix*-dominated epilithic blue-green algae may be appreciable on the shores where they occur. It must be pointed out however that blue-green algae are not always as common on the supralittoral fringe of the seashore as they are in this particular habitat.

The sand-dune slack region studied is a brackish habitat and as such is relevant to this paper. The slack is generally moist and occasionally becomes flooded. Among the higher plant biomass there is a rich growth of *Nostoc* particularly in the spring and in late autumn. *Nostoc* also occurs abundantly in summer in moist areas which are shaded from the desiccating effects of the sun's rays. Nitrogen fixation was followed *in situ* both in open areas and in the more shaded areas. The seasonal variations noted are shown in Figure 1. Little fixation occurred during the winter months and in August fixation also dropped off markedly in the open areas of the slack. In the more shaded areas fixation was particularly high in August and September,

Table VII Nitrogen fixation on a monthly basis on a *Calothrix*-dominated supralittoral fringe of a temperate seashore (after Stewart, 1967)

	μg N present per 5.0 cm² sample	Mean increase in μg N during preceding period	Mean μg N fixed during preceding period	Mean N fixed as % of increase in total N during same period
1965, 15 March	3,125			
24 April	4,228	1,103	333	30.2
23 May	4,015		392	
19 June	2,175		204	
15 July	234		41	
20 Aug.	368	134	36	26.9
30 Sept.	574	206	76	36.9
28 Oct.	3,018	2,444	93	3.8
8 Nov.	3,516	498	33	6.6
6 Dec.	3,825	309	25	8.1
1966, 14 Jan.	4,053	228	12	5.3
26 Febr.	4,799	746	35	4.7
24 March	5,197	398	61	15.3

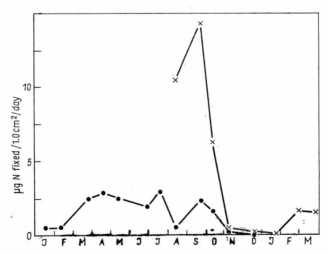

Figure 1 Seasonal variation in the *in situ* nitrogen fixation rates in open areas (●—●) and shaded areas ×—× of a sand-dune slack region. The experimental period ranged from January 1965 to March, 1966 (after Stewart, 1967)

due to the relatively high temperature and the presence of adequate moisture. Because of the variation in the degree of plant cover it was not possible to calculate reliably the amounts of nitrogen fixed over large areas but it could be calculated that in the open areas of the slack the nitrogen fixed per annum accounted for about one-fifth of the mean total nitrogen present while in the damper areas it was considerably higher, particularly in the summer.

FACTORS GOVERNING NITROGENASE ACTIVITY IN THE SEA

The quantitative significance of nitrogen fixation in the sea depends not only on the appropriate organisms being present but also on whether the environmental conditions are suitable for the synthesis of the enzyme and for its subsequent functioning. It is not the purpose of this paper to consider the factors which affect metabolism in general but only to consider those which affect nitrogenase activity. Three in particular merit special attention. These are iron, molybdenum and combined nitrogen.

Iron and molybdenum are essential components of the nitrogenase enzyme which can be separated into two major fractions neither of which fix nitrogen alone but which do so on combination. One is an iron-molybdenum protein with a molecular weight near to 100,000–135,000 and with one or two molybdenum atoms and about 16 iron atoms per mole. The other is an iron protein with a molecular weight near to 40,000–50,000 and with 3 iron atoms per mole (Burris, 1969).

Iron is one of the more abundant minor elements in the sea. Even so, as in other ecosystems, the controlling factor is not the total iron present but rather its availability. Soluble iron is used most easily by marine micro-organisms but because of the high salt content of sea water, the relatively high pH and the aeration of the upper waters there is very little iron in solution. Goldberg (1955) records that the average total iron content of sea water is 0.01 mg/l and it is known that nitrogen-fixing micro-organisms require at least ten times this amount for optimum nitrogen fixation. Thus, even if all the iron present was available for nitrogen fixation it would still not support optimum nitrogen fixation rates. Menzel and Ryther (1962) studied the availability of various nutrients, including iron, in the Atlantic and found using ^{14}C uptake as a measure of productivity that iron was the major limiting factor in their experiments, although if iron was added in excess nitrogen and phosphorus rapidly became limiting.

Molybdenum shortage does not appear to be a problem in the sea, because the average molybdenum concentration in sea water of 0.01 mg/l (Goldberg, 1955) is optimal for nitrogen fixation by blue-green algae (Arnon, 1958; Okuda, Yamaguchi and Nioh, 1962) and Menzel and Ryther (1961) obtained no stimulation of ^{14}C uptake when molybdenum was added to phosphorus- and nitrogen-supplemented Sargasso Sea waters. Stewart (1964) observed a stimulation of nitrogenase activity by *Calothrix* when he added a mixture of trace elements including iron and molybdenum to natural sea water. However, Allen (1963) observed an inhibition of fixation by a different species of *Calothrix* when nutrients including iron, molybdenum and phosphorus were added. The differences are perhaps attributable to the different concentrations of nutrients used in the two studies. The overall conclusion is that shortage of iron could be responsible in part for the paucity of nitrogen-fixing plants in marine ecosystems.

There is a wide literature on the effects of combined nitrogen on biological nitrogen fixation, which if taken at face value, give very divergent views (see Perminova, 1968). As Stewart (1969) points out in relation to the reported inhibition of nitrogen fixation by combined nitrogen "what has not always been appreciated is that inhibition does not always occur and that the degree and type of inhibition depend on the level of combined nitrogen supplied". In general combined nitrogen exerts two effects. First, it may inhibit synthesis of the enzyme and second it may affect the functioning of pre-formed nitrogenase. The data of Strandberg and Wilson (1968) emphasise the position for heterotrophic nitrogen-fixing micro-organisms. They showed that cell-free extracts of *Azotobacter* fix nitrogen at ammonium-nitrogen concentrations of up to 40 mg/l and that 150 mg/l inhibits the synthesis of the nitrogenase enzyme. Similarly with blue-green algae, Stewart *et al.* (1968) showed that 100 mg nitrate-nitrogen/l did not inhibit the activity of preformed nitrogenase in *Nostoc muscorum*, although it inhibited nitrogenase synthesis. Munson and Burris (1969) further showed, using *Rhodospirillum rubrum*, that cells grown in the presence of 100 mg/l of ammonium-nitrogen and in the absence of N_2 fixed nitrogen vigorously when subsequently exposed to N_2 in the absence of ammonium-nitrogen. From these findings they conclude that so long as there is a shortage of nitrogen for growth nitrogenase will be synthesised.

One must thus consider whether the types and levels of combined nitrogen which occur in the sea are likely to affect nitrogenase activity. The level of ammonium-nitrogen is generally higher than that of nitrate-nitrogen in

16*

surface waters, often by about two to four times (see Menzel and Spaeth, 1962 and Vaccaro, 1965), and the importance of ammonia can be guaged from the data of Dugdale and Goering (1967) who found that the uptake of nitrate-nitrogen as a percentage of ammonia plus nitrate-nitrogen ranged from 8.3 per cent in subtropical waters to 39.5 per cent in the northern Atlantic. The data of Cooper (1933) for the English Channel and of Menzel and Spaeth (1962) for the tropical Atlantic give an indication of the levels of ammonium-nitrogen in the sea. Cooper considered that in the open water the upper ammonium-nitrogen limit was 30 µg/l although in inshore waters levels of up to 180 µg/l may occur as a result of drainage from land. Menzel and Spaeth (1962) found that in general levels were highest in summer and lowest in winter with a maximum of 19 µg ammonium-nitrogen/l in October. The nitrate plus nitrite-nitrogen levels recorded by Cooper (1933) were as high as 120 µg/l in winter but this decreased to negligible quantities in summer, while Riley (1957) found that levels of approximately 7.0 µg/l of nitrate- plus nitrite-nitrogen were usually present in the surface waters of the Sargasso Sea. When these levels are considered in relation to nitrogen fixation it seems that they will seldom, if ever, inhibit the functioning of pre-formed nitrogenase and if the data of Munson and Burris (1969) are characteristic they are unlikely to inhibit nitrogenase synthesis in the sea either. These suppositions are supported in fact by field data which show that in natural ecosystems potential nitrogen-fixing micro-organisms usually do fix nitrogen if they are metabolically active.

THE PRODUCTION OF EXTRACELLULAR PRODUCTS BY MARINE NITROGEN-FIXING ORGANISMS

Nitrogen fixation is important because it is a source of *new* nitrogen to the ecosystem and of the three major ways in which it may enter the marine nitrogen cycle: cell autolysis, grazing by marine animals, and the production of extracellular products by the nitrogen-fixing organisms, the latter is perhaps the most interesting from a physiological and ecological viewpoint. The topic of algal extracellular products in general has been considered recently by Fogg (1966).

Jones and Stewart (1969 a) studied the production of extracellular nitrogen by the marine *Calothrix scopulorum* and found in the laboratory that a considerable proportion of the nitrogen fixed was liberated in to the medium. They found that the quantities of extracellular nitrogen produced were highest immediately on transfer of the alga from one set of environmental conditions to another and that the more sub-optimum the new conditions were the more extracellular nitrogen which was liberated. They concluded that on the seashore where environmental conditions change rapidly as a result of such factors as tidal movement, salinity and temperature, over 64 per cent of the nitrogen fixed by *Calothrix* may be liberated extracellularly. On the basis that *Calothrix* fixes 2.5 g $N/m^2/ann.$ (Stewart, 1967) one can calculate that at least 1.8 g $N/m^2/ann.$ will be liberated into the sea. This may be of considerable significance in intertidal habitats, and oceanic waters also if *Trichodesmium* liberates similar quantities.

The extracellular products of *Calothrix* like other nitrogen-fixing blue-green algae (Watanabe, 1951; Fogg, 1952; Whitton, 1965) are principally bound amino acids or peptides, and traces of free amino acids (Stewart, 1963; Jones and Stewart, 1969*a*) which may become available for the growth of associated organisms either directly or after breakdown. Proteolytic bacteria which convert peptides to amino acids and deaminating bacteria which reduce the amino acids to ammonia are both abundant in marine inshore waters (Zobell, 1963) and have also been noted during (Waksman, Stokes and Butler, 1937; Johnston, 1955) or just after (Duursma, 1961) plankton blooms. The ability of some algae to utilise organic material directly may be advantageous to them under nitrogen-limiting conditions but few prefer organic nitrogen to inorganic forms. Examples of marine algae which use some types of organic nitrogen are *Porphyra tenera* (Iwasaki, 1967), *Nannochloris, Stichococcus* (Ryther, 1954), *Hymenomonas, Syracosphaera, Ochrosphaera* (Pintner and Provasoli, 1963), *Hemiselmis* (Droop, 1957), *Melosira* and *Coscinodiscus* (Guillard, 1963). The latter author concluded from a survey of marine phytoplankton, that most of the algae tested grew poorly on organic nitrogen.

Jones and Stewart (1969*b*) studied the uptake of ^{15}N-labelled extracellular products of *Calothrix* by other organisms and found that all species tested including salt marsh fungi, Chlorophyta, Phaeophyta and Rhodophyta (Table VIII) showed uptake and in the case of *Chlorella marina* at least these extracellular products acted as the sole nitrogen source for growth, although growth was much slower than on inorganic nitrogen. A considerable proportion of the nitrogen removed by the various test organisms resulted from adsorption rather than active uptake (Table IX) and adsorption onto inorganic nuclei also occurred (Table X). It seems that if our findings for the extracellular products of *Calothrix* are characteristic the organic nitrogen

Table VIII Marine organisms which became labelled with ^{15}N when incubated in the presence of ^{15}N-labelled extracellular products of *Calothrix scopulorum* (after Jones and Stewart, 1969*a*, *b*)

Organism	Organism
Bacteria	Algae (contd.)
Mixed culture	*Enteromorpha* sp.
Fungi	*Enteromorpha intestinalis*
Dendryphiella salina	*Fucus spiralis*
Stachybotrys atra	*Pelvetia canaliculata*
Trichothecium roseum	*Porphyra umbilicalis*
Algae	*Scytosiphon lomentaria*
Chlorella marina	*Synechocystis* sp.
Cladophora rupestris	*Ulothrix* sp.
Dunaliella tertiolecta	*Ulva* sp.

which is found in solution in sea water probably represents only a small proportion of that available for metabolism by marine organisms and that the major proportion of the extracellular nitrogenous products liberated by marine organisms is removed fairly rapidly from solution by active uptake and particularly by adsorption.

Table IX Uptake of ^{15}N-labelled extracellular products of *Calothrix scopulorum* by living and dead samples of marine micro-organisms (after Jones and Stewart, 1969*b*)

Organism		Mean atom% Excess ^{15}N
Dendryphiella salina	Living	1.970
	Dead	0.218
Trichothecium roseum	Living	0.734
	Dead	0.197
Mixed bacteria	Living	0.133
	Dead	0.086
Cladophora rupestris	Living	0.845
	Dead	0.378

The extracellular products contained 0.93 mg nitrogen/l labelled with 4.055 atom% excess ^{15}N.

Table X Adsorption of ^{15}N-labelled extracellular products of *Calothrix scopulorum* on to inorganic nuclei (after Jones and Stewart, 1969 b).

Experiment	μg N per sample	μg N in aggregates	Mean atom % excess ^{15}N in aggregates	N in aggregates as % of total N
1	203	66	1.000	33
2	203	63	1.218	32

The extracellular nitrogen was labelled with 4.510 atom per cent excess ^{15}N and the inorganic nuclei were obtained by increasing the pH of the artificial marine medium to approximately pH 10.0.

SUMMARY

The ^{15}N method and the acetylene reduction technique are the most suitable for measuring nitrogen fixation in the sea. To date there is evidence of nitrogen fixation for 18 marine blue-green algae. These belong to the genera *Anabaena*, *Calothrix*, *Microchaete*, *Nodularia*, *Nostoc*, *Rivularia* and *Trichodesmium*. There is no evidence that any of the larger algae belonging to the Chlorophyta, Phaeophyta and Rhodophyta fix nitrogen. Potential nitrogen-fixing photosynthetic bacteria are common in shallow marine sediments and marine heterotrophic bacteria which may fix nitrogen are strains of *Azotobacter*, *Clostridium*, *Desulfovibrio*, *Mycobacterium*, *Spirillum* and *Pseudomonas*. The evidence for nitrogen-fixation by marine yeasts is not convincing. Certain marine lichens which contain blue-green algae as phycobionts also fix nitrogen. There are probably other marine nitrogen-fixing organisms which have not yet been discovered.

Quantitative *in situ* data on marine nitrogen fixation are available only for *Trichodesmium*, *Calothrix* and *Nostoc*. *Trichodesmium* can grow as well in tropical waters on N_2 as it can on combined nitrogen, while on the supralittoral fringe of temperate shores dominated by *Calothrix* 2.5 g nitrogen/ m^2/ann. may be fixed. About 1.8 g of this may be liberated extracellularly by the algae and there is evidence that a large proportion of it is removed from solution by active uptake and by adsorption.

Of the factors governing nitrogen fixation in the sea, shortage of energy probably limits the development of heterotrophic nitrogen fixing microorganisms while shortage of iron may limit the development of nitrogen-fixing organisms in general. The levels of combined nitrogen which occur in the sea are unlikely to markedly inhibit nitrogen fixation.

Acknowledgements

Much of my own work reported in this paper has been made possible through grants from the Royal Society, the Natural Environment Research Council and the Science Research Council.

References

ALLEN, M. B. (1963). Nitrogen fixing organisms in the sea. In *Marine Microbiology* (ed. C. H. Oppenheimer), 85–92. Springfield, Illinois: C. C. Thomas.

ARNON, D. I. (1958). The role of micronutrients in plant nutrition with special reference to photosynthesis and nitrogen assimilation. In *Trace Elements* (ed. C. A. Lamb, O. G. Bentley and J. M. Beattie), 1–32. New York: Academic Press.

BENECKE, W. and KEUTNER, J. (1903). Über stickstoffbindende Bakterien aus der Ostsee. *Ber. dtsch. botan. Ges.* **21**, 333–345.

BERNARD, F. and LEÇAL, J. (1960). Plançton unicellulaire recolté dans l'ocean Indian per le Charcot (1950) et le Norsel (1955–1956). *Bull. Instit. Oceanogr. Monaco*, No. 1166.

BIGGINS, D. R. and POSTGATE, J. R. (1969). Nitrogen fixation by cultures and cell-free extracts of *Mycobacterium flavum* 301. *J. gen. Microbiol.*, **56**, 181–193.

BREMNER, J. M., CHENG, H. H. and EDWARDS, A. P. (1965). Assumptions and errors in nitrogen-15 tracer research. *Rep. FAO/IAEA Tech. Meeting Brunswick-Volkenrode*, 9–14 *Sept.*, 1963. 429–442. Oxford: Pergamon Press.

BRONGERSMA-SANDERS, M. (1948). The importance of upwelling water to vertebrate paleontology and oil geology. *Verhandel. Koninkl. Ned. Akad. Wetenschap., Afdel. Natuurk.*, Sect. II. **45**, 1–112.

BRONGERSMA-SANDERS, M. (1957). Mass mortality in the sea. In *Treatise on marine ecology and paleoecology*, Vol. I., *Ecology* (ed. J. W. Hedgpeth, 941–1010), Geological Soc. Amer. Mem. No. 67.

BURRIS, R. H. (1969). Progress in the biochemistry of nitrogen fixation. *Proc. Roy. Soc.* B **172**, 317–347.

BURRIS, R. H., EPPLING, F. J., WAHLIN, H. B. and WILSON, P. W. (1943). Detection of nitrogen fixation with isotopic nitrogen *J. Biol. Chem.*, **148**, 349–357.

BURRIS, R. H. and WILSON, P. W. (1957). Methods for measurement of nitrogen fixation. In *Methods in Enzymology* (ed. S. Kolowick and N. O. Kaplan), **4**, 355–365. New York: Academic Press.

COOPER, L. H. N. (1933). Chemical constituents of biological importance in the English Channel. Pt. I. Phosphate, silicate, nitrate, nitrite, ammonia. *J. Mar. Biol. Assoc. U. K.*, **18**, 677–728.

DELSMAN, H. C. (1939). Preliminary plankton investigations in the Java Sea. *Treubia*, **17**, 139–181.

DILWORTH, M. J. (1966). Acetylene reduction by nitrogen-fixing preparations from *Clostridium pasteurianum. Biochim. Biophys. Acta*, **127**, 285–294.

DROOP, M. R. (1957). Auxotrophy and organic compounds in the nutrition of marine phytoplankton. *J. gen. Microbiol.*, **16**, 286–293.

DUGDALE, R. C. and GOERING, J. J. (1967). Uptake of new and regenerated forms of nitrogen in primary productivity. *Limnol. Oceanogr.*, **12**, 196–206.

DUGDALE, R. C., MENZEL, D. W. and RYTHER, J. H. (1961). Nitrogen fixation in the Sargasso Sea. *Deep-Sea Res.*, **7**, 298–300.

DUGDALE, R. C., GOERING, J. J. and RYTHER, J. H. (1964). High nitrogen fixation rates in the Sargasso Sea and the Arabian Sea. *Limnol. Oceanogr.*, **9**, 507–510.

DUURSMA, E. K. (1961). Dissolved organic carbon, nitrogen and phosphorus in the sea. *Netherlands J. Mar. Res.*, **1**, 1–148.

FARRAN, G. P. (1932). The occurrence of *Trichodesmium thiebautii* off the south coast of Ireland. *Rapp. Proc. Verb. Cons. Int. Explor. Mer.* **77**, 60–64.

FEDOROV, M. V. and KALININSKAYA, T. A. (1961 *a*). New forms of nitrogen-fixing organisms isolated from soddy-podzolic soils. *Dokl. mosk. sel'-khoz. Akad. K. A. Timiryazeva* **70**, 145.

FEDOROV, M. V. and KALININSKAYA, T. A. (1961 *b*). A new species of a nitrogen-fixing *Mycobacterium* and its physiological peculiarities. *Mikrobiologiya*, **30**, 9.

FEDOROV, M. V. and KALININSKAYA, T. A. (1961 *c*). The relation of the nitrogen-fixing mycobacterium (*Mycobacterium* sp. 301) to various carbon sources and to additional growth factors. *Mikrobiologiya*, **30**, 833.

FELDMAN, J. (1932). Sur la biologie des *Trichodesmium* Ehrenberg. *Rev. Algol.*, **6**, 357–358.

FOGG, G. E. (1952). The production of extracellular nitrogenous substances by a blue-green alga. *Proc. Roy. Soc. B.* **139**, 372–397.

FOGG, G. E. (1956). Nitrogen fixation by photosynthetic organisms. *Ann. Rev. Plant Physiol.*, **7**, 51–70.

FOGG, G. E. (1966). The extracellular products of algae. *Oceanogr. Mar. Biol. Rev.*, **4**, 195–212.

FOGG, G. E. and HORNE, A. J. (1967). The determination of nitrogen fixation in aquatic environments. In *Chemical environment in the aquatic habitat*, 115–120 (ed. H. L. Golterman and R. S. Clymo), Amsterdam: N. V. Noord-Hollandsche Uitgerers Maat-schappij.

GEITLER, L. (1932). Cyanophyceae. In Rabenhorsts *Kryptogamenflora*, **14**, Leipzig.

GOLDBERG, E. D. (1955). Biogeochemistry of Trace Metals. In *Treatise on marine ecology and paleoecology*, Vol. 1, *Ecology* (ed. J. W. Hedgpeth), 345–358. Geological Soc. Amer. Mem. No. 67.

GUILLARD, R. R. L. (1963). Organic sources of nitrogen for marine centric diatoms. In *Symposium on Marine Microbiology* (ed. C. H. Oppenheimer), 93–104. Springfield, Illinois: C. C. Thomas.

HARDY, R. W. F. and BURNS, R. C. (1968). Biological nitrogen fixation. *Ann. Rev. Biochem.*, **37**, 331–358.

HARDY, R. W. F., HOLSTEN, R. D., JACKSON, E. K. and BURNS, R. C. (1968). The acetylene-ethylene assay for N_2 fixation: Laboratory and field evaluation. *Plant Physiol.*, **43**, 1185–1207.

HELLRIEGEL, H. and WILFARTH, H. (1888). Untersuchungen über die Stickstoffnahrung der Gramineen und Leguminosen. *Beilageheft zu der Zeitschrift des Vereins f. d. Rüben-zucker-Industrie d. D. R.: Berlin*, 234 pp. (This paper gives details of Hellriegel and Wilfarth's 1886–1888 work.)

HILL, S. and POSTGATE, J. R. (1969). Failure of putative nitrogen-fixing bacteria to fix nitrogen. *J. gen. Microbiol.* **58**, 277-286.

IWASAKI, H. (1967). Nutritional studies of the edible seaweed *Porphyra tenera*. II. Nutrition of *Conchocelis*. *J. Phycol.*, **3**, 30–34.

JOHNSTON, R. (1955). Biologically active compounds in the sea. *J. Mar. Biol. Ass.* U. K., **34**, 185–195.

JONES, K. and STEWART, W. D. P. (1969*a*). Nitrogen turnover in marine and brackish habitats. III. The production of extracellular nitrogen by *Calothrix scopulorum*. *J. Mar. Biol. Ass.* U. K., **49**, 475–488.

JONES, K. and STEWART, W. D. P. (1969*b*). Nitrogen turnover in marine and brackish habitats. IV. Uptake of the extracellular products of the nitrogen-fixing alga *Calothrix scopulorum*. *J. Mar. Biol. Ass.* U. K., **49**, 701–716.

KLUCAS, R. V. (1967). Ph. D. thesis, University of Wisconsin.

KOBAYASHI, M., TAKAHASHI, E. and KAWAGUCHI, K. (1967). Distribution of nitrogen-fixing micro-organisms in paddy soils of Southeast Asia. *Soil Sci.*, **104**, 113–118.

KOBAYASHI, M. (1968). Nitrogen fixation and a role of photosynthetic bacteria. *Proc. 1st Symp. of Nitrogen fixation and nitrogen cycle, Tokyo, Dec.* 19–20, 45–46.

KONDRAT'EVA, E. N. (1965). *Photosynthetic bacteria*, 243 pp. Jerusalem: Israel Program For Scientific Translations.

KRISS, A. E., MISHUSTINA, I. E., MITSKEVICH, N. and ZEMTSOVA, E. V. (1967). *Microbial population of oceans and seas.* 287 pp. London: Edward Arnold.

MENZEL, D. W. and RYTHER, J. H. (1961). Nutrients limiting the production of phytoplankton in the Sargasso Sea, with special reference to iron. *Deep Sea Res.*, **7**, 276–281.

MENZEL, D. W. and SPAETH, J. P. (1962). Occurrence of ammonia in Sargasso Sea waters and in rain water at Bermuda. *Limnol. Oceanogr.*, **7**, 159-162.

MILLBANK, J. W. (1969). Nitrogen fixation in moulds and yeasts—a reappraisal. *Arch. Mikrobiol.*, **68**, 32–39.

MUNSON, T. O. and BURRIS, R. H. (1969). Nitrogen fixation by *Rhodospirillum rubrum* grown in nitrogen-limited continuous culture. *J. Bacter.*, **97**, 1093–1098.

NEESS, J. C., DUGDALE, R. C., DUGDALE, V. A. and GOERING, J. J. (1962). Nitrogen metabolism in lakes. I. Measurement of nitrogen fixation with N^{15}. *Limnol. Oceanogr.*, **7**, 163–169.

OKUDA, A., YAMAGUCHI, M. and NIOH, I. (1962). Nitrogen fixing micro-organisms in paddy soils. X. Effect of molybdenum on the growth and the nitrogen assimilation of *Tolypothrix tenuis*. *Soil. Sci. Plant. Nutrit.*, **8**, 35–39.

PARKE, M. and DIXON, P. S. (1968). Check-list of British marine algae—Second revision. *J. Mar. Biol. Ass.* U. K. **48**, 783–832.

PERMINOVA, G. A. (1968). Growth of nitrogen-forming blue-green algae in the presence of bound nitrogen. *Microbiology*, **37**, 551–554.

PINTNER, I. J. and PROVASOLI, L. (1958). Artificial cultivation of a red pigmented marine blue-green alga *Phormidium persicinum*. *J. gen. Microbiol.*, **18**, 190–197.

PINTNER, I. J. and PROVASOLI, L. (1963). Nutritional characteristics of some chrysomonads. In *Symposium on Marine Microbiology* (ed. C. H. Oppenheimer), 114–121, Springfield, Illinois: C. C. Thomas.

PSHENIN, L. N. (1963). Distribution and ecology of *Azotobacter* in the Black Sea. In *Marine Microbiology* (ed. C. H. Oppenheimer), 383–391. Springfield, Illinois: C. C. Thomas.

RAMAMURTHY, V. D. and KRISHNAMURTHY, S. (1968). Nitrogen fixation by the blue-green alga *Trichodesmium erythraeum* (Ehr.) *Curr. Sci.*, **37**, 21–22.

REINKE, J. (1903). Die zur Ernährung der Meeres-Organismen disponiblen Quellen an Stickstoff. *Ber. dtsch. bot. Ges.* **21**, 371.

RILEY, G. A. (1957). Phytoplankton of the north central Sargasso Sea. *Limnol. Oceanogr.*, **2**, 252–270.

RYTHER, J. H. (1954). The ecology of phytoplankton blooms in Moriches Bay and Great South Bay, Long Island, New York. *Biol. Bull., mar. Biol. Lab., Woods Hole*, **106**, 198–204.

SCHÖLLHORN, R. and BURRIS, R. H. (1966). Study of intermediates in nitrogen fixation. *Fed. Proc.*, **25**, 710.

SCHÖLLHORN, R. and BURRIS, R. H. (1967). Acetylene as a competitive inhibitor of nitrogen fixation. *Proc. nat. Acad. Sci.*, **58**, 213–216.

SISLER, F. D. and ZOBELL, C. E. (1951). Nitrogen fixation by sulfate reducing bacteria indicated by nitrogen/argon ratios. *Science*, **113**, 511–516.

STEWART, W. D. P. (1963). Liberation of extracellular nitrogen by two nitrogen-fixing blue-green algae. *Nature, Lond.*, **200**, 1020–1021.

STEWART, W. D. P. (1964). Nitrogen fixation by Myxophyceae from marine environments. *J. gen. Microbiol.*, **36**, 415–422.

STEWART, W. D. P. (1965). Nitrogen turnover in marine and brackish habitats. I: Nitrogen fixation. *Ann. Bot.* N. S., **29**, 229–239.

STEWART, W. D. P. (1966). *Nitrogen fixation in plants*, 168 pp. London: Athlone Press of the University of London.

STEWART, W. D. P. (1967). Nitrogen turnover in marine and brackish habitats. II: Use of ^{15}N in measuring nitrogen fixation in the field. *Ann. Bot.*, N. S. **31**, 383–407.

STEWART, W. D. P. (1968). Nitrogen input into aquatic ecosystems. In *Algae, Man and the Environment* (ed. D. F. Jackson), 53–72. Syracuse: Syracuse University Press.

STEWART, W. D. P. (1969). Biological and ecological aspects of nitrogen fixation by free-living micro-organisms. *Proc. Roy. Soc.* B. **172**, 367–388.

STEWART, W. D. P., FITZGERALD, G. P. and BURRIS, R. H. (1967). *In situ* studies on N_2-fixation using the acetylene reduction technique. *Proc. Nat. Acad. Sci.*, **58**, 2071–2078.

STEWART, W. D. P., FITZGERALD, G. P. and BURRIS, R. H. (1968). Acetylene reduction by nitrogen-fixing blue-green algae. *Arch. Mikrobiol.*, **62**, 336–348.

STRANDBERG, G. W. and WILSON, P. W. (1968). Formation of the nitrogen-fixing enzyme system in *Azotobacter vinelandii*. *Canad. J. Microbiol.*, **14**, 25–31.

VACCARO, R. F. (1965). *Inorganic nitrogen in sea water*. In *Chemical Oceanography*, Vol. I (ed. J. P. Riley and G. Skirrow), 365–408. London: Academic Press.

VAN BAALEN, C. (1962). Studies on marine blue-green algae. *Botan. Mar.*, **4**, 129–139.

WAKSMAN, S. A., HOTCHKISS, M. and CAREY, C. L. (1933). Marine bacteria and their role in the cycle of life in the sea. II. Bacteria concerned in the cycle of nitrogen in the sea. *Biol. Bull., mar. Biol. Lab., Woods Hole*, **65**, 137–167.

WAKSMAN, S. A., STOKES, J. L. and BUTLER, M. R. (1937). Relation of bacteria to diatoms in sea water. *J. mar. biol. Ass.* U. K. **22**, 359–373.

WATANABE, A. (1951). Production in cultural solution of some amino acids by the atmospheric nitrogen-fixing blue-green algae. *Arch. Biochem. Biophys.* **34**, 50–55.

WHITTON, B. A. (1965). Extracellular products of blue-green algae. *J. gen. Microbiol.*, **40**, 1–11.

WILSON, P. W. (1958). Asymbiotic nitrogen fixation. In *Encyclopaedia of Plant Physiology*, **8**, 9–47. Berlin: Springer.

WOOD, E. J. F. (1965). *Marine microbial ecology*, 243. London: Chapman and Hall Ltd.

ZOBELL, C. E. (1946). *Marine microbiology*, 233. Waltham, Mass: Chronica Botanica Co.

ZOBELL, C. E. (1963). Domain of the marine microbiologist. In *Marine microbiology* (ed. C. H. Oppenheimer), 3–24, Springfield, Illinois: C. C. Thomas.

Ocean circulation in monsoon areas

BRUCE TAFT

Scripps Institution of Oceanography
University of California
La Jolla, California

Abstract

The near-surface circulation of the northern Indian Ocean is primarily wind-driven. In the central part of the ocean there is a lag between change of phase of the monsoon and change in surface circulation of about one month. In the Arabian Sea the response of the boundary currents to the change in monsoon phase is more complex and there is not a simple lag time. Some recent results suggest that a zonal band of vortices may extend eastward from the Somali Coast during both monsoon phases and that continuous zonal currents, such as are shown on the average surface current charts, do not exist. Because of the monsoon winds the surface temperature distributions in the Arabian Sea change markedly within the year. Coastal upwelling of subsurface water is most pronounced during the southwest monsoon and is most intense on the western side of the sea rather than the eastern side. Upwelling north of the Somali Current and off the southwest coast of India does not appear to be primarily wind-driven coastal upwelling. Cool water at the surface is due to the dynamic response of the subsurface density distribution to the set up of the coastal currents. Off the Arabian coast the cool water at the surface is due to wind-driven upwelling. There is no upwelling of cool water at the equator in the Indian Ocean during either monsoon phase. This appears to be due to the presence of eastward wind stress at the equator in the Indian Ocean rather than westward stress. The Ekman theory of wind-driven currents would predict an equatorial convergence in the Indian Ocean.

There is some variation in the use of the word *monsoon* in the climatological literature. The word has been used consistently to describe a wind system with a significant seasonal change. Different criteria have been used to define a significant seasonal change: some writers have required a complete seasonal reversal of winds; other have required only a seasonal deviation from the predominant wind direction during the year.

Flöhn (1960)* has published a world map which shows the areas with large changes in wind direction during the year and the constancy of the wind in these areas (Fig. 1). Regions where the resultant wind direction varies during the year by more than 120° is shown by shading. Differences in shading indicate the constancy of the wind. Over the Indian Ocean north of 10° S, and extending into the western Pacific, there is a region where the winds change by more than 120° and where the winds also have a constancy greater than 60 per cent.

Resultant wind directions and wind constancy for the months of January and July are represented in Figures 2a and 2b. In January wind fields of the Pacific and Indian oceans are similar to each other: both oceans have a northeast wind north of the equator and southeast trades south of the equator. The primary difference between the two oceans is that in the Indian Ocean the northeast monsoon extends across the equator and the Intertropical Convergence Zone, where the northeast monsoon and the southeast trades meet, lies south of the equator; in the Pacific Ocean the Intertropical Convergence Zone, where the northeast and southeast trades meet, lies north of the equator. In July the winds are entirely different in the Indian Ocean. The southeast trades blow across the equator and the wind vector north of the equator rotates clockwise. Over the Arabian Sea and the Bay of Bengal the winds are southwesterly.

In the Indian Ocean the timing of the monsoon periods is remarkably symmetrical. The strongest winds of the northeast monsoon are in January and the strongest winds of the southwest monsoon are in July; each monsoon persists for five months. Wind speeds during the southwest monsoon are considerably stronger than during the northeast monsoon.

Average surface current vectors for January, the month of strongest winds during the northeast monsoon, are shown in Figure 3a. In the central Indian Ocean there are two westward-flowing surface currents with an

* The map is a reproduction of a map drawn by S. P. Chromow. Flöhn's reference to the paper by Chromow could not be verified so a reference to the original paper is not given.

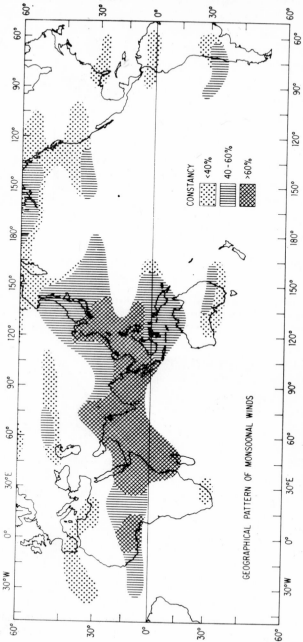

Figure 1 Regions where the resultant wind changes by more than 120° during the year are shown by shading. Constancy of the wind in these regions is indicated by differences in shading. The chart is from Flöhn (1960)

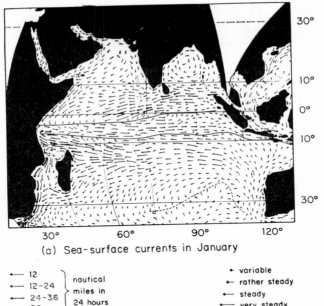

(a) Sea-surface currents in January

← 12	} nautical
← 12-24	miles in
← 24-36	} 24 hours
← 36	

← variable
← rather steady
← steady
← very steady

—— marked limits between currents flowing in different directions or
limits between different water types (lines of convergence & divergence)

- - - - other often weakly developed limits mostly between subtropical or
tropical waters

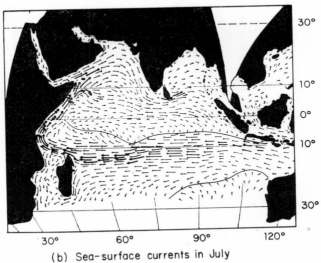

(b) Sea-surface currents in July

Figure 2a, b Average sea-surface currents for (a) January and (b) July.
Constancy and speed of the current are indicated by the length and width
of the arrows. The charts are from Defant (1961)

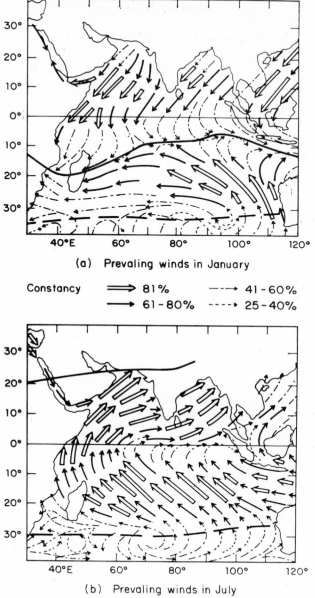

(a) Prevaling winds in January

Constancy   ⟹ 81%   ⤑ 41-60%

⟶ 61-80%   ⤍ 25-40%

(b) Prevaling winds in July

Figure 3a, b   Prevailing surface winds in (a) January and (b) July. Direction of wind is based on dominant wind directions computed for each 5-degree square. Constancy of the wind is indicated by the type of arrow. The charts are from U.S. Weather Bureau (1938) eastward-flowing Equatorial Countercurrent lying between them. Off the Somali coast the Somali Current flows southwestward with moderate speeds. There are weak currents to the southwest off the coast of Arabia and weak currents to the southeast off the southwest coast of India.

The circulation is vastly different during the southwest monsoon (Fig. 3b). In July the westward-flowing Northeast Monsoon Current has been replaced by the eastward-flowing Southwest Monsoon Current, so that the surface currents north of 5° S are all directed eastward and the boundaries of the Equatorial Countercurrent become unclear. The Somali Current flows northwestward as an intense western boundary current to about 8° N. At 8° N the Somali Current leaves the coast and turns eastward. There is a suggestion of an anticyclonic turning of the current where it leaves the coast. Off the Arabian coast the currents are weak and directed northwestward. Off the southwest coast of India the current is southeastward as it was in January.

There is little doubt that the circulation of the surface layer of the Indian Ocean is primarily wind-driven. Comparison of the monthly wind and current charts indicates that the change from the westward-flowing Northeast Monsoon Current to the eastward-flowing Southwest Monsoon Current north of the equator takes place about one month after the monsoon winds have reversed. The set-up of the currents along the boundaries of the Arabian Sea appears to be complex and there is not a simple lag time between current and wind. For example, the average current charts show that the Somali Current in April extends northward to 8° N but the southeast trades only reach as far north as the equator in April. The northward-flowing Somali Current is present during seven months of the year even though each phase of the monsoon lasts five months.

Off the west coast of India there is even a larger discrepancy between winds and currents. Currents are southward or southeastward during eight months and northwestward during two months. The northwestward current flows opposite to the wind. The circulation along the eastern boundary of the Arabian Sea must be driven in part by thermocline forces.

Average surface current charts give the impression that the currents change from one quasi-stationary state to another in response to the change in monsoon phase. Currents on these charts appear to be coherent large-scale flows but these charts show the monthly currents averaged over a large number of years. Figure 4 is one of a series of charts published by Düing (in press) showing the dynamic height of the sea surface relative to 800 db in the Arabian Sea. The hydrographic observations are from the months of July, August and September in 1963. As would be expected, the Somali Current is shown by the gradient in dynamic height off the Somali coast. In addition, there is a pattern of alternating highs and lows in dynamic height east of the Somali Current. These highs and lows may represent a

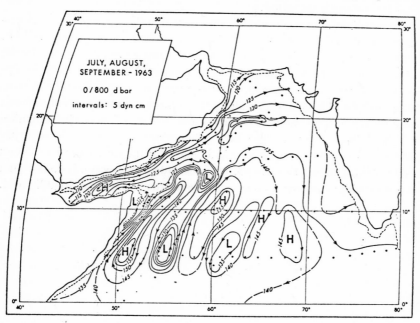

Figure 4 Dynamic height of the sea-surface relative to 800 db for July, August and September of 1963. The chart is from Düing (in press)

row of anticyclonic and cyclonic vortices. According to Düing, who has examined a large number of such charts, similar patterns persist in recognizable form through both monsoon phases. It is possible that the highs and lows merely represent the effect of internal waves but the amplitudes of the waves that would be required to account for the pattern is larger than seems reasonable (Düing, in press). At his point the existence of the pattern of vortices must be considered speculative. However, if the vortices are real then mixing of the near-surface waters would be promoted and the rate of nutrient enrichment of the surface layer of the Arabian Sea would be increased.

17*

Distributions of temperature and nutrient salts in the Arabian Sea are affected by the monsoon wind changes. The focus of my discussion will be on the distribution of sea-surface temperature because it is probably the best indicator available of vertical motion or vertical mixing which will bring nutrient salts into the euphotic zone where they can be utilized by plants in primary production.

Because of the monsoon winds in the Arabian Sea, there is a large difference between the distribution of surface temperature in the Arabian Sea and the surface temperature distribution in the north Atlantic and north Pacific oceans. In the Atlantic and Pacific oceans there is an intense western boundary current and a broad, slowly-moving eastern boundary current with upwelling of cooler water along the eastern boundary (Wooster and Reid, 1963). Concentrations of nutrients and the rate of primary production are higher along the eastern boundary than along the western boundary.

The average surface temperature in January for the western Indian Ocean north of 5° S is shown in Figure 5. The west side of the ocean is cooler than the eastern side; isotherms slope southwestwards toward the Somali coast.

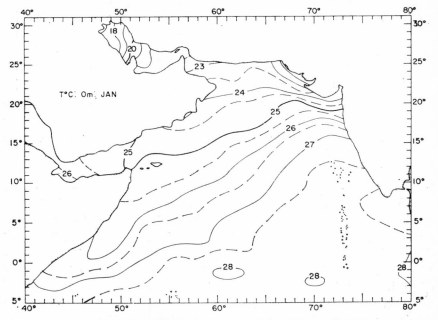

Figure 5   Average sea-surface temperature (°C) for the month of January.
The chart is from Wooster *et al.* (1967)

There is a temperature minimum off the Somali coast which is presumably connected with the advection of cooler water to the south by the south-westward flow of the Somali Current. Temperatures off the Arabian and Somali coasts are everywhere above 24°C. There are no coastal regions with low temperatures which suggest upwelling—except for a small region off the Pakistan coast. The Ekman theory of wind-driven currents would call for upwelling of cool water off the southwest coast of India because of the component of the wind parallel to this coast (Smith, 1968). However, there is only a slight southward trend of the isotherms along the coast; up-welling is much more pronounced when the southwest monsoon is blowing. The lack of upwelling along the southwest coast of India may be due to the relatively low wind speeds of the northeast monsoon. The patch of cooler water off Pakistan does not seem to be a clear case for wind-induced upwelling because at this time the winds are blowing almost directly offshore (Fig. 2a).

The dramatic change in surface temperature brought about by the change in monsoons is evident in the average surface temperature chart for August (Fig. 6). Because of the increased radiation during the summer months there

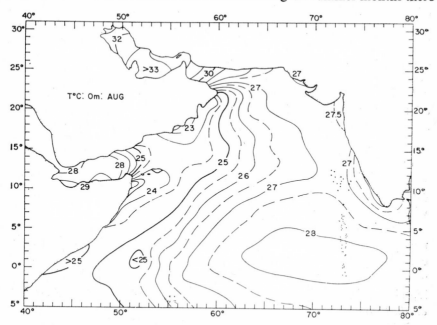

Figure 6   Average sea-surface temperature (°C) for the month of August.
The chart is from Wooster *et al.* (1967)

is a substantial increase in surface temperature everywhere east of 60° E—except off the southwest coast of India. However, cooler water is found off the coasts of Somalia and Arabia—there are isolated patches of water with temperatures less than 22°C.

The low temperatures off the Somali coast occur north of where the Somali Current turns eastward abruptly and leaves the coast. Temperatures in this region can be considerably lower than indicated in Figure 6. On the ARGO survey of August–September 1964, temperatures less than 14°C were measured along with very low values of oxygen saturation (Stommel and Wooster, 1965).

Winds are parallel to the Somali coast and the Ekman theory would predict upwelling along the western boundary. There is little evidence of upwelling of water along the coast in the region of the Somali Current itself, i.e., south of 8° N. This is probably due to the large horizontal velocity gradient across the current which would tend to promote the mixing of the upwelled water so that it would not be recognizable. North of the Somali Current this would not be the case; one would expect the effects of the upwelling to be fairly well preserved. Wind-induced upwelling is at least partially responsible for the occurrence of the cold water region shown in Figure 6. The abrupt temperature decrease suggests that there is also another cause. Because the cross-stream balance of forces in a western boundary current is essentially geostrophic, isotherms in the Somali Current north of the equator will slope up toward the coast. If the volume transport of the current is to remain constant, then the slope of the isotherms across the current must increase to the north because of the increase of the Coriolis parameter. Isotherms on the inshore side of the Somali Current do ascend downstream and the thermocline appears to break the surface at the cold water region (Warren *et al.*, 1966). The "outcrop" of the thermocline is probably the result of a dynamic constraint on the current and not due to wind-induced upwelling.

The cold water region off the Arabian coast is most certainly a region of upwelling set up by the strong winds parallel to the coast. The surface currents here are not particularly strong and the southwest winds must drive water offshore and set up coastal upwelling. This upwelled water is very high in nutrients and supports a high level of primary productivity (Wooster *et al.*, 1967).

Off the southwest coast of India there is a region of cooler water with temperatures less than 26°C. Currents are directed southward along the

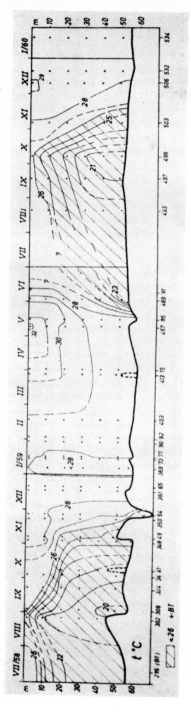

Figure 7 Vertical temperature (°C) distribution at position B off Cochin, India (July 1958 to January 1960). Open circles and broken bottom contours indicate stations near position B. The figure is from Banse (1968)

coast and the winds are essentially westerly (Fig. 2b and 3b). At the southern end of India the wind has a slight northerly component. Because the winds do not have a strong component parallel to the coast the seasonal appearance of this cool water along the coast must be due to some other cause.

Banse (1968) has published a time series of the vertical temperature distribution in the shallow water off Cochin on the southwest coast of India (Fig. 7). Temperatures less than 26°C begin to appear in June at the beginning of the southwest monsoon and persist until November when the transition between the southwest monsoon and the northeast monsoon takes place. The change in temperature at 60 m exceeds 8°C during both 1958 and 1959. Maximum development of the cold water at this location approximately coincides with the maximum strength of the southeastward current along the coast. Nutrient and phytoplankton concentrations are relatively high along this coast during the southwest monsoon and the higher concentrations are surely related to the seasonal upwelling of cooler water which is shown in the time series (Banse, 1968).

Banse (1968) has suggested that the appearance of the cold water is not primarily due to wind-induced upwelling, but results from the tilting of the thermocline up toward the coast to provide a geostrophic cross-current balance of forces in the southeastward coastal current. Isotherms will have to slope up toward the coast if the current is to be in geostrophic balance, and the slope will increase as the strength of the current increases. It is remarkable that there is such a short time constant for the adjustment of the field of mass to the change in winds.

One of the major results of the International Indian Ocean Expedition was the description of the Equatorial Undercurrent of the Indian Ocean. The undercurrent, which is located in the thermocline as it is in the other two oceans, is not present at all times of the year. Direct measurements of the velocity profile at the equator are summarized in Figure 8. Measurements that show an Undercurrent are indicated by boxes; the numbers in the boxes give the current speed. A circle indicates no Undercurrent. There are obvious gaps in the distribution of observations: there are no observations in October and November and only one each in December and January. Most of the observations that showed an Undercurrent were made during March, April and May (Swallow, 1964; Taft and Knauss, 1967). The absence of the Undercurrent in December, January and on one of two February measurements suggests that the Undercurrent is set up at the end of the northeast monsoon. The time lag between the set up of the westward-flowing

Northeast Monsoon Current at the surface and the eastward-flowing Under-current in the thermocline is of the order of three months.

The average surface temperature for the world ocean in February is shown in Figure 9. In the Pacific there is a region of minimum temperature along the equator. Between longitudes 110° and 130° W the minimum is

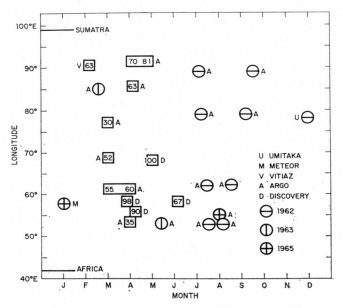

Figure 8 Direct measurements of current velocity at the equator in the Indian Ocean. Circles represent current profiles that did not show an Equatorial Undercurrent. Squares represent a measurement of the Under-current with the maximum speed of the Undercurrent in cm/sec given by the numbers in the squares. Ships and year of observation are indicated in the legend. All Undercurrent measurements were made in 1963 and 1964

isolated; therefore, the low temperatures cannot be related to westward advection of cold water from the Peru Current. Associated with the low temperatures are high concentrations of nutrients and high levels of primary production (Wyrtki, 1966). Note that neither the Indian or the Atlantic oceans have an equatorial minimum during February.

The usual explanation of the equatorial minimum in the Pacific is that the westward stress of the trade winds stets up a divergence of the horizontal velocity field in the surface layer at the equator and upwelling of water from

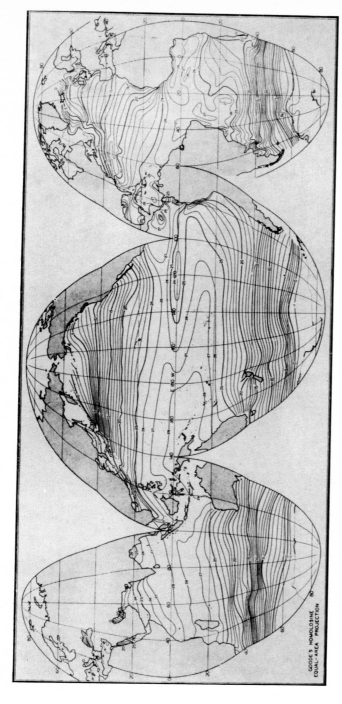

Figure 9 Average sea-surface temperature (°C) for the world ocean in February. The chart is from Sverdrup *et al.* (1942)

the thermocline must occur in order to satisfy the conservation of mass (Smith, 1968). The annual mean zonal component of wind stress at the equator in the Pacific is given in Figure 10. The mean wind stress between longitudes 110° and 160° W has a strong westward component. Horizontal lines above and below the mean values give the seasonal range of the stress; only at the eastern and western margins do they overlap zero.

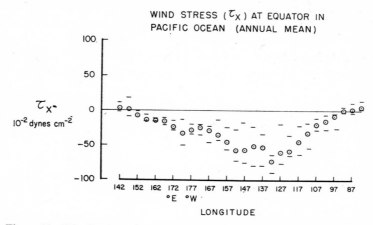

Figure 10 Distribution of annual average zonal wind-stress component (10^{-2} dyne/cm^2) along equator in Pacific Ocean. Range of average quarterly zonal stresses is indicated by the horizontal lines. Positive stress is eastward and negative stress is westward. Data is from Hidaka (1958)

There is another possibility that might explain the observed distribution of surface temperature. Because of the Equatorial Undercurrent in the upper thermocline in the Pacific, there is a very large vertical shear in the mean current which might be expected to induce a large amount of vertical mixing between waters in the surface layer and the upper thermocline. Vertical mixing would produce the observed distribution of surface temperature; the width of the temperature minimum and the width of the Undercurrent are approximately equal. Because the dynamics of mixing processes in the ocean are poorly understood, it has not been possible to rule out vertical mixing and attach primary importance to wind-induced upwelling in the maintenance of the equatorial minimum.

If the minimum were due to vertical mixing it might be expected to be present in the Indian Ocean at the end of the northeast monsoon when the Undercurrent is present (Fig. 8). However, vertical sections across the equator in the Indian Ocean do not show the equatorial minimum in surface temperature (Taft and Knauss, 1967). The mean vertical shear of velocity in the Indian Undercurrent is roughly comparable to that in the Pacific Undercurrent. Although the Pacific Equatorial Undercurrent is stronger than the Indian Equatorial Undercurrent it is also thicker so that the vertical shear above the velocity maximum is approximately the same. The conclusion is that mixing is probably not primarily responsible for the equatorial minimum of sea-surface temperature in the Pacific and that the winds are the decisive factor.

In Figure 11 is plotted the annual mean zonal component of wind stress along the equator in the Indian Ocean. The magnitude of the zonal wind stress is

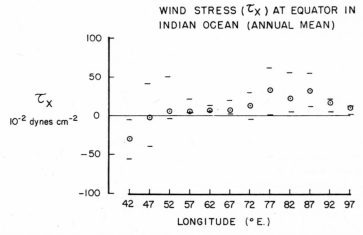

Figure 11 Distribution of annual average zonal wind-stress component (10^{-2} dyne/cm^2) along equator in Indian Ocean. Range of average quarterly zonal stresses is indicated by the horizontal lines. Positive stress is eastward and negative stress is westward. Data is from Hidaka (1958)

smaller in the Indian Ocean than in the Pacific and it tends to be eastward. According to Ekman theory, this would call for equatorial convergence not divergence, and upwelling would not occur. The lack of an equatorial minimum in surface temperature and the low nutrient concentrations along the equator in the Indian Ocean during both monsoon phases is probably due to the distribution of wind stress on the equator. Zonal wind stress components are low and are eastward in direction so that wind-driven upwelling at the equator would not be expected in the Indian Ocean.

Acknowledgements

This work has been supported by the Office of Naval Research, under contract NONR 2216 (23), and by the Marine Life Research Program of the Scripps Institution of Oceanography.

References

BANSE, K. (1968). Hydrography of the Arabian Sea shelf of India and Pakistan and effects on demersal fishes. *Deep-Sea Res.* **15**, 45–79.

DEFANT, A. (1961). *Physical Oceanography.* Vol. I. London: Pergamon.

DÜING, W. (in press). The monsoon regime of the currents in the Indian Ocean. International Indian Ocean Expedition Oceanographic Monographs. No. 1. Honolulu: East-West Center Press.

FLÖHN, H. (1960). Monsoon winds and general circulation. In *Monsoons of the World.* pp. 65–74. New Delhi: Hind Union Press.

HIDAKA, K. (1958). Computation of the wind stress over the oceans. *Rec. Oceang. Wks. Japan,* **4**, 77–123.

SMITH, R. L. (1968). Upwelling. *Oceanogr. Mar. Biol. Ann. Rev.,* **6**, 11–46.

STOMMEL, H. and WOOSTER, W. S. (1965). Reconnaissance of the Somali Current during the southwest monsoon. *Proc. Nat. Acad. Sci.,* **54**, 8–13.

SVERDRUP, H. U., JOHNSON, M. W. and FLEMING, R. H. (1942). *The Oceans.* New York: Prentice Hall.

SWALLOW, J. C. (1964). Equatorial Undercurrent in the western Indian Ocean. *Nature,* **204**, 436–437.

TAFT, B. A. and KNAUSS, J. A. (1967). The Equatorial Undercurrent of the Indian Ocean as observed by the Lusiad expedition. *Bull. Scripps. Inst. Oceanogr.,* **9**, 1–163.

U. S. WEATHER BUREAU (1938). *Atlas of Climatic Charts of the Oceans.* Weather Bureau No. 1247. Washington: Govt. Printing Office.

WARREN, B., STOMMEL, H. and SWALLOW, J. C. (1966). Water Masses and patterns of flow in the Somali Basin during the southwest monsoon of 1964. *Deep-Sea Res.,* **13**, 861–888.

WOOSTER, W. S. and REID, J. L., Jr. (1963). Eastern boundary currents. In *The Sea,* pp. 253–280 (ed. Hill, M. N.). London: Interscience.

WOOSTER, W. S., SCHAEFER, M. B. and ROBINSON, M. K. (1967). *Atlas of the Arabian Sea for Fishery Oceanography.* IMR Ref. 67–12. La Jolla: Univ. Calif. Inst. Mar. Resources.

WYRTKI, K. (1966). Oceanography of the eastern equatorial Pacific Ocean. *Oceanogr. Mar. Biol. Ann. Rev.,* **4**, 33–68.

The echinoderms of the Ilha Grande region (RJ, Brasil)*
Distribution and abundance of six species up to the isobath of 50 m

LUIZ ROBERTO TOMMASI

Instituto Oceanográfico da Universidade de São Paulo, Brasil

Resumo

No presente trabalho, são apresentadas algumas observações sôbre a distribuição de seis espécies de equinodermes na região de Ilha Grande, Estado do Rio de Janeiro (SP, Brasil), Foi verificado que a região apresenta áreas ricas e áreas pobres em equinodermes. As áreas ricas provàvelmente estão relacionadas com a presença de densos bancos de lamelibrânquios, no caso de asteróides, e com o enriquecimento da água pela mistura de água de baixa salinidade mas rica em fitoplancton, partículas orgânicas em suspensão vinda da baía de Sepetiba com água de maior salinidade porém menos produtiva vinda da plataforma continental. Foi verificada uma relação entre a presença de água subtropical (formada pela mistura de água da corrente do Brasil, com das Malvinas e que se desloca para o norte, sob a corrente do Brasil) e espécies subantárticas e antárticas de ofiuróides. Verificou-se também uma nítida separação de nichos ecológicos entre duas espécies de holothurioides. A distribuição das seis espécies estudadas é do tipo agregado.

* This paper was partially supported by the São Paulo State Research Foundation.

I INTRODUCTION

A "Meeting on the Natural History of Aquatic Organisms" took place in 1963, sponsored by the São Paulo State Research Foundation, aiming at the compilation of bibliography, exchange of information on the inves- tigations under way, and the determination on a priority basis of areas for investigation.

As a result of this meeting, in 1965–1966 the São Paulo State Research Foundation sponsored a survey of the marine fauna and flora of the Ilha Grande region (RJ) from the coast down, to 50 m deep. The survey was carried out in collaboration with the Department of Zoology, São Paulo State Department of Agriculture, and with the Department of Zoology and Department of Botany, University of São Paulo. The region was chosen due to its 388 isles, 2,000 beaches, 7 bays, and several inlets, presenting a great variety of ecological environments and a rich fauna and flora. It is a region of geographical transition (Vannucci, 1964; Tommasi, 1967) and may be reached easily by sea and land.

The Instituto Oceanografico da Universidade de São Paulo was in charge of the gatherings on the soft bottom of the region, dredgings, sediment and water sampling. The sampling continued until March 1969 after the program had been finished (Map 1).

The present paper gives some observation on six species of echinoderms and characteristic of different environmental conditions of the region.

II SPECIES STUDIED

1 Cucumaria manuelina Tommasi (in press)

The *Cucumaria* species Blainville, 1834 occur generally on soft bottom. According to Ursin (1960) *Cucumaria elongata* Düben and Koren, occurs mainly in the region of mixture of central water of the North Sea (char- acterized by *Sagitta setosa*) with water from the north of the North Sea (characterized by *Sagitta elegans*), on muddy bottom. Such distribution pattern has also been observed for *Acrocnida brachiata* (Montagu) and *Echinocardium pennatificium* Norman. The geographic distribution of *Cucumaria elongata* Düben and Koren is similar to that of *Cucumaria hyndmani* (Thompson) (from Norway to the Mediterranean Sea), however, the latter occurs on sandy and detrital bottom. This preference for different types of sediments is probably responsible for the smaller penetration of the species into the North Sea (Ursin, 1960).

Cucumaria manuelina occurred, except for station 330, only the region in front of Ilha Grande, stations 308, 313 as far as station 288, in front of Restinga da Marambaia and in the region between Ilha Grande and the entrance of Sepetiba Bay, up to the line formed by stations 201 and 172. This occurrence is very similar to that of *Cucumaria hyndmani* (Thompson). The species occurred from 15 m down to 50 m, with bottom water temperature ranging from 19.20 to 22.4°C and salinity from 34.07 to 36.02°/oo. In 49.95 % of the stations where it has occurred the prevailing granulometric interval of the sediment was 1000–420 μ. The sediment median was between 44–760 μ, whereas at most stations it was higher than 500 μ.

The higher densities of the species were recorded in the region near Ilha Pau a Pino (St. 303, 304), on rich calcareous bottom (fragments of brachiopod *Bouchardia rosea* shells).

2 Protankira benedeni (Ludwig, 1881)

Protankira benedeni Deichmann, 1930, p. 309–310; Tommasi, 1969, p. 17, fig. 26.

This species was collected at depths varying from 2.5 m to 26 m deep, bottom water salinity from 30.25°/oo to 35.87°/oo and temperature from 22.08 to 28.00°C.

In 67.32% of the stations where this species was recorded the granulometric fractions prevailing was 44 μ. The median was below 44 μ to 260 μ. In fraction 44 μ it showed a clear trend to be more frequent on less calcareous sediments.

3 Astropecten armatus braziliensis Mull. et Troschel, 1842

Astropecten braziliensis braziliensis, Döderlein, 1917, p. 83, 169. Est. I, VIII.

Astropecten armatus braziliensis, Tortonese, 1956, p. 326, fig. 2; Tommasi, 1958, p. 12, est. 11, fig. 3.

This species was collected at depths from 2 m to 50 m bottom water temperature from 16.2 to 25.75°C, and salinity ranging from 32.10 to 35.60°/oo (72% of the stations from 35–35.9°/oo). The sediment median of the stations where it occurred was from 1750 to 44 μ. In 58.52% of the stations the prevailing granulometric fraction was 420–44 μ.

4   Hemipholis elongata (Say, 1825)

Hemipholis elongata Clark, 1933, p. 46; Thomas, 1962, p. 686–689, fig. 22. It is this ophiuroid which occurred more densily in the region, namely, 1069 specimens/m². (St. 182). It occurred in 143 stations, being collected at depths varying from 3 to 27 m, with water temperature ranging from 16.55 to 26.00°C and salinity from 32.10 to 35.75⁰/₀₀. According to Ferguson (1948) the lowest salinity tolerance limit is 23⁰/₀₀. At. 42.60% of the stations in which this species was recorded the prevailing granulometric fraction was 44 μ.

5   Amphiura kinbergi Ljungman, 1871

Amphiura kinbergi Thomas, 1965, p. 638–641, fig. 1.
Amphiura kinbergiensis Koehler, 1914, p. 52, Est. IV, fig. 3–4, Est 5, Fig. 1–2.
This species was collected at depths varying from 9 to 42 m, with bottom water temperature from 16.5 to 26.70°C, and salinity from 29.66⁰/₀₀ to 35.59⁰/₀₀. At. 55.55% of the stations where this species occurred the prevailing granulometric fraction of the sediment was 44 μ.

6   Amphiura joubini (Koehler, 1912)

Amphiura joubini Mortensen, 1936, p. 277–279, figs. 16–17; Tommasi, 1967 b, p. 3–5, fig. 2.
Amphiura joubini Fell, 1961, p. 40–43, pl. 17, fig. 4.
At. 33.33% of the stations where this species occurred the prevailing granulometric fraction of the sediment was 420–149 μ. The median was 44 to 470 μ, being lower than 100 μ at most stations. It has been collected at depths varying from 9 to 49 m, temperature 16.1 to 26.70°C, and salinity from 33.40 to 36.04⁰/₀₀. According to Fell (1961), in the Atlantic it is a species indicating muddy and sandy bottom, a statement which agrees with the sedimentological analysis of the stations under observation.

DISCUSSION

The most important feature of a sedimentary substratum is to maintain approximately steady the chemical conditions of the bottom and of the water immediately adjacent, forming a complex where chemical and biochemical processes take place, interrelated with biological processes (Magliocca and Kutner, 1965). The size of the sediment grains act on the benthic fauna, directly as well as indirectly. Some of the processes related to the grains size are the following; method and capture of food, fixation, movements. The organic content and the size of grains are related with the speed of the current that passes on the sediment. The same is true for its aerobic and anaerobic nature (Craig and Jones, 1966). Stable sediments are propitious for fixation or displacements of many benthonic forms while very fluid sediments (denoting the size of the grains and the sediment porosity, apud Weller, 1959) or yet very thick or unstable sediments, are unfavorable for the fixation of many species (Craig and Jones, 1966). The biomass is generally low in very fine mud and in very well sorted sand and can reach maximum values in muddy sand and in bottoms with detritus containing shell fragments. Fine sand is susceptible to erosion and has low organic matter content. However, if the organic matter content increases, a parallel increase in the fine fractions takes place along with an increase in the resistence of the sediment towards erosion, since organic matter and mud aggregate the grains. Consequently, the sediment gets somewhat rich, with ideal characteristic for benthic life (Craig and Jones, 1966). The greater the finer sediment fraction, the richer it will be in organic matter (Zenkevitch, 1963). Due to the small renovation of the water layer on this type of bottom, the oxygen content is low. This fact is intensified by the bacterial activity which determines the formation of substances toxic to animal life, such as chloridric acid: hence, the occurrence of bottom with very low biomass. These muds are dark colored and stink (Theede, Ponat, Hiroki and Schlieper, 1969). The presence of currents with low dissolved oxygen content causes the formation of similar bottoms (Galhardo, 1963). A turbulent flow determines an erosion more intense than a laminar flow. The occurrence of pebbles and many pelecypods valves, for instance, may transform a laminar flood into a turbulent flood and result in a stronger tendency of the sediment to erode (Menard and Boucot, 1951; Johnson, 1957). The sediment characteristic is determined basically by three fundamental processes: the settling velocity, the roughness velocity, and the threshold velocity (Inman, 1949; Sanders, 1958). The grains more easily displaced by the currents are those of 0.18 mm diameter. If a sediment is constituted mainly by grains of that size, this would mean the occurrence of a small erosion activity and therefore the presence of an environment stable enough and highly favorable to benthic organisms (Sanders, 1958). According to Boillot (1964) there is a close correlation between the quantity of calcareous material in the sediment and richness of the benthic fauna. The thick banks of *Ophiothrix* and of bryozoa (i.e. *Cellaria*) are recorded in sediments with more than 50% calcareous material (zoogenos sediment). Driscoll (1967) verified that several filtering forms of the epifauna (*Astrangia, Microciona, Anadara, Anomia, Crepidula, Chaetopleura*) occurred mainly on bottoms rich with shell fragments and exposed to strong currents. There are benthic association that, due to being related to edaphic factors (currents, pollution, turbidity; Pérès, 1967), may occur at different environments. That is the case, for instance, with the associations of sand and pebble placed under the deep currents and which are characterized by the presence of *Amphioxus* (Pérès, 1961). The relation between the occurrence of benthic species and the extant hydrologic conditions has been stressed by Ursin (1960), for instance, and by Rasmussen (1965) who showed that in the North Sea *Henricia* occurs in waters rich in nutrients, characterized by *Sagitta elegans*. Blacker (1957) showed that several echinoderm species, among other animals, might be considered as indicators of hydrographic changes in the region of Spitsbergen. Thorson (1961, 1966) showed the importance of the currents on the distribution of larvae of some benthic species. Ursin (1960) showed the presence of areas rich and poor in echinoids. According to Ursin (1960), the rich areas are apparently in the region of transition between two water masses. According to McIntosh (1967) distribution may be defined as being the area occupied by a species in which it is established; abundance, as the quantity of individuals of a certain species in the various parts of its distribution area. A species is homogeneously distributed in a certain region when the number of individuals is constant in all parts of the region, or else, when propability of being found there is unvariable (McIntosh, 1961).

Areas rich and poor in echinoderm species

The polychaetes were the predominant animals in 22 out of the 46 quantitative stations carried out with the vessel "*Emilia*"; the lamellibranchs, especially *Nucula semiornata**, were predominant in 8 stations; the crustaceans, especially the amphypods were numerous in 10 stations; the cephalochordates (*Branchiostoma platae*) in 3 stations and the brachiopods (*Bouchardia rosea*) in 2 stations. The echinoderms were predominant in only 6 stations (St. 21, 28, 30, 39, 181 and 182). The more important echinoderms in soft bottoms are the ophiuroids. In a general way, the most important group is that of the polychaetes, followed by the ophiuroids and the lamellibranchs.

---

* Determined by Miss Lucia Penna, Zoological Museum, U. S. P.

It is important to point out that in the stations near the Ilha da Gipoia, i.e., in the region where banks of *Astropecten armatus braziliensis* were found, lamellibranchs were dense.

Table 1 gives the stations made on this area.

Table 1 Density of animals from stations in the Ilha da Gipoia

Station number	No. of Polychaetes m²	No. of Lamellibranchs m²	No. of Crustaceans m²	No. of Echinoderms m²	Other m²
93	6	54	24	6	18
99	145.2	2,927.1	36.3	103.5	1,165
100	409.2	2,623.5	244.2	337.3	806
111	354	1,026	132	· 84	510

At stations 99 and 100 the high density of lamellibranchs are due to *Nucula semiornata*; at Station 99 the density of the other benthic groups is due to the great number of *Discoporella umbellata* (bryozoa); at station 100 this group has also showed high density due to *Edwardsia* sp. (anthozoa), *Dentalium gouldi* (Scaphopod) and *Discoporella unbellata*.

The poorest regions in echinoderms species were those of Parati, Parati-Mirim, Mamanguá and the inner part of Sepetiba Bay:

Station: 7, 8, 10, 13, 59, 75, 85, 106, 120, 130, 150, 151, 158, 159, 161, 162, 178, 185, 191, 193, 196, 200, 201, 202, 204, 205, 219, 225, 226, 232, 238, 257, 261, 265, 276, 277, 278, 279, 293, 323, 324, 337, 365, 367.

At the following stations, no benthic organism has been recorded although sediment has been collected:

Station: 133, 141, 198, 222, 287.

The reason for the absence of echinoderms may be due to two distinct factors:

a. the regions of low salinity with small water circulation and fine sediment, sometimes rich in material being decomposed; that is the case, for instance, with the Enseada de Parati, Parati-Mirim, Mamanguá and Sepetiba Bay;

b. regions exposed to strong hydrodynamics, especially waves, currents, etc. That is the case with stations located near the outer part of the Ilha Grande and between the Ilha de Itacurussa and the continent.

18*

Ursin (1960) mentions that there are areas rich and poor in echinoderms. According to the same author, apparently the richest areas are those located in the zones of transition between two water masses. He verified that the waters of the Baltic Sea show low salinity, yet they are very productive in phytoplankton, and are rich in nutrients. The oceanic waters show higher salinity, being therefore more favourable to echinoderms; they are, however, poorer in nutrients than those of the Baltic Sea. When the two water masses mix they form a very rich environment. The environmental effects may be noted mainly at localities where the water movement is swift, allowing a quantity of useful substances to get continually in contact with the animals (Ursin, 1960). Probably the same is true for the Sepetiba Bay. The water in that Bay presents low salinity, yet, as it has been observed, is very rich in phytoplankton. It is possible that in the region located on the west of Ilha Grande represents a mixture of shelf water with coastal water from the Ribeira, Jacuacanga and other entrances.

The inner regions, located southwesterly of the Ilha Grande Bay (region of the Aricó and Parati entrances) and at the Sepetiba Bay show a progressing decrease in salinity towards the center. It seems that in those regions the mixing processes mentioned previously are not present.

The abundance of echinoderms in the region

At a first glance, the ophiuroid populations would lead us to think of temporary concentration around pole of attraction (food, for instance) of individuals normally scattered in a certain area. However, various authors (Thorson, 1957; Barnard and Ziesenhenne, 1961; Cabioch, 1961; Reese, 1966) recorded a steady location of those banks at different regions.

There is a specific difference between the ophiuroid bottoms of region such as Roscoff, Mediterranean, Gullmar Fjord and those found at the region under study, and also, for instance, at the southern coast of California. At those two latter regions the ophiuroids recorded are *Amphiuridae* family, while at the others they are Ophiotrichidae. The Amphiuridae live generally hidden in the sediment, with only their arms out (Buchanan, 1964); the Ophiotrichidae comprise epifauna species, being vagile and not sedentary as the Amphiuridae. However both feed mainly on organic matter suspended or deposited on the bottom. They are heterotrophous of the detritus food chain and play an important role as decomposers in the ocean trophic cycle.

According to Barnard and Ziesenhenne (1960) the predominance of ophiuroids in a community is the result of palezoogeographic, hydro-climatic factors, sediment characteristics and biological competition. Different species of various genera predominante on different bottoms in the same region. The *Ophiothrix* banks occur on pebbles bottoms, with stable and uniform epifauna (Cabioch, 1961), while those of *Amphiura-Amphiodia-Amphioplus* are associated with sand bottoms, muddy-sand bottoms (Thorson, 1957). Hence, the *Ophiothrix* banks are associated with strong wave currents that hinder the deposition of fine particles and the formation of muddy bottoms (Pruvet, 1897; Boillot, 1960). On the other hand, the amphiuridea bottoms are always associated with muddy bottoms (Thorson, 1957). Such type of bottoms occur at regions with smaller hydrodynamics which permits the deposition of very fine particles (Thorson, 1957).

At the area under study, three regions were found with a great abundancy of echinoderms, namely:

1) Region of Ilha da Gipoia
2) West region near Ilha Grande
3) Region of the Sepetiba Bay entrance.

According to Cabioch (1961) the presence of thick banks of ophiuroids is probably associated with stable food conditions. The tide currents play an important role in the transportation and sedimentation of particles useful as food and consequently in the distribution of ophiuroid banks. At sandy regions in front of Ilha Grande, Restinga da Marambaia and between the Ilha Grande and Ilha da Marambaia, *Cucumaria manuelina* occurred. On the other hand, *Protankira benedeni*, occurred in muddy regions, mainly at the Saco de Mamanguá, Parati-Mirim, Baía da Ribeira, Baía de Jacuacanga and Baía de Sepetiba. Its major abundance was observed at Station 183, where 34 specimens were found. The sandy bottoms where *Cucumaria manuelina* occurred and in which we found a great number of brachiopods (*Bouchardia rosea*) and of cephalochords (*Branchiostoma platae*) suggest an association of these species with places of strong currents. The muddy bottoms where *Protankira benedeni* occurred show its association with places of deposition.

Distribution of Amphiura joubini (Koehler) and of Amphiura kinbergi Ljungman

Amphiura joubini is a eurythermic and euryhalinic species which occurrs principally in the west region of the Ilha Grande and in the shelf in front of the region presently studied. This species occurs in the subantarctic and antarctic region and is carried to the north by the subtropical water mass under the Brazilian current (Tommasi, 1967). As was shown by Emilsson (1963) in different areas of the littoral region, this water mass penetrated into bays and inlets, to little depth. This is the same situation in the west region of the Ilha Grande. Where this water mass occurred, we found *Amphiura joubini*, and believe that this species is an indication of the presence of the subtropical water mass (temperature of 10–20°C and salinity of 35–36°/oo).

Amphiura kinbergi occurred principally in the northwestern region of the Ilha Grande and in the entrance of the Sepetiba Bay. While *Amphiura joubini* is typically a shelf species, which penetrates in the coastal region carried by the subtropical water mass, *Amphiura kinbergi* is a species of the coastal region, occurring also in the region under the influence between the two species:

	Temperature	Salinity	Median
Amphiura kinbergi	16.5–26.70°C	31.38–35.59°/oo	420–149 μ
Amphiura joubini	16.1–26.70°C	33.40–36.04°/oo	44 μ

Homogeneity of the distribution

The distribution of the six echinoderm species discussed in paper, and analysed by the graphic distribution of frequences, is of the aggregate type of distribution.

Similar results were obtained by Buchanan (1967) with five echinoderms species of Scotland and by Ursin (1960) with echinoderm of the North Sea central region. Ladd (1957) showed that the great majority of the echinoderms of the recent seas are gregarious and that this type of life relates to ancient geological times and the history of this group.

References

BARNARD, J. L. and ZIESENHENNE, F. C. (1961). Ophiuroids communities of Southern California Coastal bottoms. *Pacific Naturalist*. **2**, 132–152.

BLACKER, R. W. (1957). Benthic animals as indicator of hydrographic conditions and climatic conditions and climatic change in Svalbard waters. *Fish. Inv. Ser.* II, 20–49.

BOILLOT, G. (1960). La repartition des fonds sous-marins au large des Roscoff. *Cah. Biol. Mar. Paris*. **1**, 3–23. (1964). Géologie de la Manche Occidentale. *Ann. Inst. Ocean.* **42**, 220.

BUCHANAN, J. B. (1964). A comparative study of some features of the biology of *Amphiura filiformis* and *Amphiura chiajei* (Ophiuroidea) considered in relation to their distribution. *J. Mar. biol. Ass. U. K.* **44**, 565–576.

CABIOCH, L. (1961). Etude de la répartition des peuplements benthiques au large de Roscoff. *Cahiers Biol. mar. Roscoff.* **II**, 1–40.

CLARK, H. L. (1933). A handbook of the littoral echinoderms of Porto Rico and the other West Indies Islands. *Bull. Sci. Surv. Porto Rico and Virgin Island.* **16**, 147.

CRAIG, G. Y. and JONES, N. S. (1966). Marine benthos, substrate and palaeoecology. *Palaeontology.* **9**, 30–38.

DEICHMANN, E. (1930). The holothurians of the western part of the Atlantic Ocean. *Bull. Mus. Comp. Zool. Harv.* **71**, 45–226.

DODERLEIN, L. (1917). Die Asteriden der Siboga—Exp. I. Die Gattung *Astropecten* und ihre Stammesgeschichte. *Siboga Exp., Monogr.* **46a**, 191 p. 17 Est.

DRISCOLL, E. G. (1967). Attached epifauna—Substrate relations. *Limnol. Oceanol.* **12**, 633–641.

EMILSSON, I. (1963). Levantamento oceanográfico meteorológico da Enseada do Mar Virado, Ubatuba, *SP. Sér. Ocean. Física.* **3**, 41–53.

FELL, H. B. (1961). The fauna of the Ross Sea. *New Zealand Ocean. Inst. Mem.* **18**, 79.

FERGUSON, A. H. (1948). Experiments on the tolerances of several marine invertebrates to reduced salinity. *Proc. La. Acad. Sci.* **2**, 16–17.

GALLARDO, A. (1963). Notas sobre la densidad de la fauna bentonica en el sublitoral del norte de Chile. *Gayana. Zoologia.* **10**, 15.

INMAN, D. L. (1949). Sorting of Sediments in the light of fluid mechanics. *J. sedim. Petrol.* **19**, 51–70.

JOHNSON, R. G. (1957). Experiments on the burial of shells. *J. Geol.* **65**, 527–535.

KOEHLER, R. (1914). A contribution to the study of ophiurans of the U. S. National Museum. *Smith. Inst. U. S. Bull. nat. Mus.* **84**, VII + 173 p.

MAGLIOCCA, A. and KUTNER, A. (1965). Sedimentos de fundo da enseada do Flamengo—Ubatuba. Contrcões Inst. Oceanogr. USP, *Oceanografia Física.* **8**, 14.

MCINTOSH, R. P. (1967). An index of diversity and the relation of certain concepts to diversity. *Ecology.* **48**, 392–404.

MENARD, H. W. and BOUCOT, A. J. (1951). Experiments on the movement of shells by water. *Am. J. Sci.* **249**, 131–151.

MORTENSEN, TH. (1936). Echinoidea and Ophiuroidea. *Discovery Reports.* **12**, 199–348.

PÉRÈS, J. M. (1967). Les biocenoses benthiques dans le systeme phytal. *Rec. Trav. St. Mar. End.* **42**, 3–113.

RASMUSSEN, B. N. (1966). On the taxonomy and biology of the north Atlantic species of the asteroid genus *Henricia Gray. Medd. Dan. Fisk. Hav.* N. S. V. 1, 157–213.

REESE, E. S. (1966). The complex behavior of echinoderms, in *Physiology of Echinoderms*, R. A. Boolootian ed., Interscience Publ. 157–218.

SANDERS, H. L. (1958). Benthic studies in Buzzards Bay. I. Animal-Sediment Relationships. *Limnol. Oceanogr.* **3**, 245–256.

THEESE, H., PONAT, A., HIROKI, K. and SCHLIEPER, C. (1969). Studies on the resistence of marine bottom invertebrates to oxygen deficiency and hydrogen sulphide. *Marine Biology.* **2**, 325–337.

THOMAS, L. (1962). The shallow water amphiurid brittlestars of Florida. *Bull. mar. Sci.* **12**, 623–694.

THORSON, G. (1957). Bottom Communities. *Geol. Soc. Amer.* **67**, 461–534.

THORSON, G. (1961). Length of pelagie larval life in marine bottoms invertebrates as related to larval transport by ocean currents. Oceanography. *Amer. Assoc. Adv. Sci. Publ.* **67**, 455–474.

THORSON, G. (1966). Some factors influencing the recruitments and establishment of marine benthic communities. *Neth. Journ. Sea Res.* **3**, 267–293.

TOMMASI, L. R. (1958). Os equinodermas do litoral de São Paulo. II. Contrcões. Inst. Oceanogr. USP., *Oceanogr. Biol.* **2**, 39.

TOMMASI, L. R. (1967). Sôbre dois amphiuridae da fauna marinha do Sul do Brasil. *Contrcões. Inst. Oceanogr. USP., sèr. Ocean. Biol.* **2**, Sp. 2 figs.

TORTONESE, E. (1956). Sul alcune specie de Astropectinidae. *Ann. Mus. Civ. St. Nat. Genova.* **68**, 319–334.

URSIN, E. (1960). A quantitative investigation of the echinoderm fauna of the Central North Sea. *Medd. Danm. Fisk-og Havunders. N. S.* **2**, 204.

WELLER, J. M. (1959). Compaction of sediments. *Bull. Am. Ass. Petrol. Geol.* **43**, 273–310.

ZENKEVITCH, L. (1963). *Biology of the seas of the USSR.* G. Allen and Unwin Ltd., London, 955 p.

Trophic chains observed in Paradise Harbor (*Antarctic Peninsula*) related to variations in the fertility of its waters

ALDO P. TOMO

Instituto Antártico Argentino

Abstract

This paper deals with the annual variation of the physical, chemical, hydrological and meteorological constants that influence the actual fertility of the sea, which in turn affects trophic chains.

Resumen

Se considera en este trabajo la variación anual de las constantes físico-químicas, hidrológicas y meteorológicas que ejercen influencia sobre la fertilidad actual del mar y ésta a su vez sobre las cadenas tróficas.

I. INTRODUCTION

The basic data for this research work were collected in the vicinity of Scientific Station Almirante Brown, in Paradise Harbor—Antarctic Peninsula, during 1967. The research was carried out based on the best known premises that solar energy, mineral nutrients and sea water are the main abiotic components of the marine ecosystem and that the flow of radiant energy that passes through such ecosystem is absorbed by phytoplankton, which feeds other organisms that intervene in the transference chains.

The following are the food chains that will be considered herein:

1) Food chain of the Blue-eyed Cormorant
(*Phalacrocorax atriceps*).

2) Food chain of the Silver Gray Petrel
(*Fulmarus glacialoides*).

3) Food chain of the Gentoo Penguin
(*Pygoscelis papua ellsworthii*).

4) Food chain of the Crabeater Seal
(*Lobodon carcinofagus*).

5) Food chain of the Fin Whale
(*Balaenoptera physalus*).

Additionally, mention is made of the possible cause of nonmigration in certain marine birds, which being linked by some chains to the fertility of the sea, do not migrate when it decreases in winter.

II. WORKING METHOD

Measurements on sea water salinity and nutrients were carried out at the Scientific Station Almirante Brown; in addition, sea water temperature was recorded at the surface three times a day at 08 : 00, 14 : 00 and 20 : 00 hs. Meteorological data were also recorded. Some observations were made on the circulation of surface waters that enter and outflow through the access channels that communicate the bay with Gerlache Strait.

The trophic chains regarding some birds and mammals in the area were studied by means of quantitative and qualitative analysis of the stomach contents in some of them, and the direct observations on the food that others were ingesting. Such contents were classified in the laboratories at the Scientific Station Almirante Brown and the Instituto Antártico Argentino. The physico-chemical results were the responsibility of the chemist, Mr. Norberto L. Bienati, and the Assistant of Geology, Mr. Rufino A. Comes. Incursions were made to the breeding areas in order to capture the specimens for analysis of stomach contents at a later time.

III. VARIATIONS IN THE PHYSICAL, CHEMICAL, METEOROLOGICAL, AND BIOLOGICAL CONDITIONS OF THE AREA AND THEIR INFLUENCE ON SEA WATER FERTILITY

It was observed that sea water temperature on the surface ranges from +2.8°C in January to −1.58°C in July. Then, it increases gradually toward January, achieving its maximum in this month. Sea water salinity on the surface ranges from 25.5⁰/₀₀ to 34.0⁰/₀₀ throughout the year.

Table 1 Monthly and annual averages of sea water temperature on surface, 1967
Paradise Harbor

Month	Temperature
January	2.092°C
February	1.012°C
March	0.651°C
April	−0.334°C
May	−1.004°C
June	−1.345°C
July	−1.374°C
August	−1.390°C
September	−0.940°C
October	−0.332°C
November	0.305°C
December	1.000°C
Annual	−0.138°C

Table 2 Salinity on surface, 1967
Paradise Harbor

Date	Salinity ⁰/₀₀	Date	Salinity ⁰/₀₀	Date	Salinity ⁰/₀₀
24/II	33.48	11/III	32.86	28/IV	33.89
3/III	33.53	25/III	33.35	2/V	34.09
4/III	33.53	28/III	33.44	9/V	33.78
7/III	33.40	15/IV	33.39	12/V	34.29
5/VI	33.26	29/VII	34.53	25/IX	33.98
6/VI	33.49	31/VII	33.33	28/IX	33.95
8/VI	33.49	4/VIII	33.64	3/X	33.78
12/VI	33.62	7/VIII	33.21	7/X	34.04
15/VI	33.57	10/VIII	33.78	11/X	33.99
19/VI	33.60	14/VIII	33.84	18/X	33.80
22/VI	33.60	17/VIII	33.80	21/X	33.96

Table 2 (*cont.*)

Date	Salinity °/oo	Date	Salinity °/oo	Date	Salinity °/oo
27/VI	33.51	21/VIII	33.78	27/X	33.84
30/VI	33.48	25/VIII	33.80	31/X	30.32
3/VII	33.55	28/VIII	33.78	7/XI	34.00
6/VII	33.62	2/IX	33.80	11/XI	33.98
10/VII	33.60	4/IX	33.84	15/XI	33.75
13/VII	33.55	9/IX	34.31	28/XI	33.73
17/VII	33.60	11/IX	33.95	14/XII	33.86
20/VII	35.73	14/IX	33.87	21/XII	33.86
22/VII	34.87	19/IX	34.05	27/XII	33.61
24/VII	34.13	22/IX	33.58		

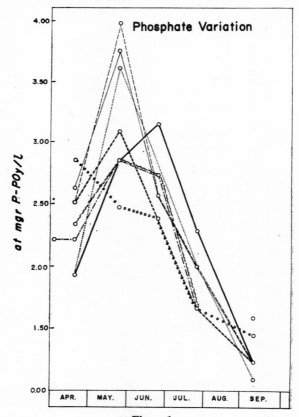

Figure 1

According to the general tendency of the curve of phosphates and nitrites values, it was observed that from April to September the concentrations for phosphates on the surface and in deep waters are near O level; then they increase, reaching a maximum in May. In June a peak occurs regarding nitrites on surface and in deep waters. The concentrations of phosphates and nitrites tend to drop to very low values at the end of winter (Figs. 1 and 2).

Luminosity presents a marked seasonal fluctuation, since a maximum occurs in December with 20 hours of direct light and a minimum on the surface in June with 4 hours of direct incident light. (It was considered that the incident direct sun light is that resulting from an angle of solar obliquity

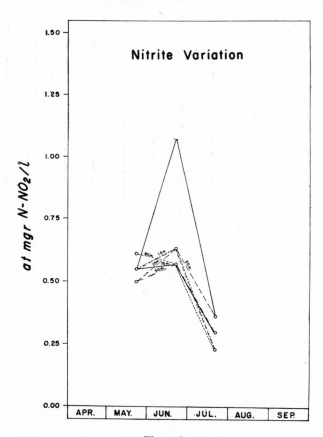

Figure 2

greater than 15°, the minimum which allows the entrance of the sun rays into the water.)

The predominant winds in the area are from the south-southwest and the north-northeast over the year. They increase in intensity during March and April and result in great turbulence in the bay's waters.

Regarding the amount of phytoplankton capable of photosyntesis, the maximum and minimum rates were not calculated since there was not a method available at that time. It could be done indirectly, however, by measuring the sea water transparency (Fig. 3). This allowed us to observe

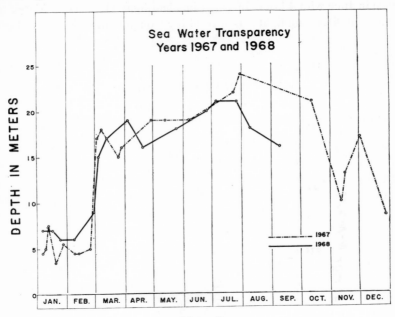

Figure 3

the fluctuation of the quantitative values of such living cells during the year, as they are restricted to the physico-chemical changes of sea water and the meterological changes of the aerial environment which influences the marine environment.

When sea water passes from the stage of potential fertility to that of actual fertility, that is, when it is able to produce organic matter, then bio-production takes place. This stage of actual fertility is observed in the bay's waters from June/July. At that time nutrients decrease remarkably due to the demand of growing phytoplankton which reach a maximum in December. This concentration is then fairly stable until March, when it falls to a minimum.

This month (March) is taken as the starting point of restoration of potential fertility. During the summer an increase in the flow of land drainage carried with thaw, results from the action of radiant energy on ice. This, plus the influence of guaneras and the dynamics of water masses, as well as that of marine plant and animal populations, sets the basis for the beginning of such mechanisms of restoration.

IV. TROPIC CHAINS OBSERVED

The following food chains were observed in the bay's waters in relation to the bioproduction.

1. Food chain of the Blue-eyed Cormorant *(Phalacrocorax atriceps)*

This marine bird that produces guano lives in the area the whole year. It breeds in groups called cormoraneras which are a source of nitrites and phosphates owing to the guano that is the product of their dejections. During the summer, it feeds on fish *(Notothenoids)*. In autumn, winter, and spring it is nourished by seaweeds. As it can be noted, this bird feeds mainly on two kinds of organisms regarding marine bioproduction.

In summer: Zooplankton *(Euphausia sp.)* ⟶ fish *(Trematomus newnesi)* ⟶ Blue-eyed Cormorant.

In winter: Seaweeds *(Gracilaria simplex* and others undetermined) ⟶ Blue-eyed Cormorant.

2. Food chain of the Gentoo Penguin *(Pygoscelis papua ellsworthii)*

The Gentoo Penguin arrives at the end of spring and starts to breed in the rookeries of the surrounding areas. Its arrival coincides with the outburst of phytoplankton, utilized by the grazing of microzooplankton which in turn are consumed by krill *(Euphausia sp.)* which is the main food of the Gentoo Penguin. Two kinds of feeding habits were observed by the analysis of the bird's stomach content, according to the abundance of the different organisms on which it feeds. One is constitued by cephalopods (undeter-

mined due to their stage of decomposition) and young fish larvae
(*Pleuragramma antarcticum*); while the other is exclusively krill (Euphausids).

    1) Cephalopods ———→ fish (*Pleurogramma antarcticum*) ———→ Gentoo
       Penguin.

    2) Krill (*Euphausia sp.*) ———→ Gentoo Penguin.

3. Food chain of the Silver Gray Petrel (*Fulmarus glacialoides*)

This migrating bird is seen in large flocks on the bay during March and
April. In these months, the medusae are abundant in sea water of that area
and the petrels feed mainly on them. When the medusae are lacking, the
gray petrels migrate and they are not seen again until the cycle is repeated.

    Medusae (undetermined) ———→ Silver Gray Petrel

4. Food chain of the Crabeater Seal (*Lobodon carcinofagus*)

This species does not live permanently in the area. It appears in groups of
less than 100 individuals which feed on krill during February and March.
Once their season of abundant feeding is over, they migrate to the north of
the Antarctic Peninsula, possibly following the northwardly drift of waters
which contain food.

    Krill (*Euphausia sp.*) ———→ Crabeater Seal.

5. Food chain of the Fin Whale (*Balaenoptera physalus*)

This mammal is seen in the bay's waters during the summer, generally
alone, rarely in groups of two. It can be mentioned that exceptionally an
individual was sighted in June. Usually, the mammals were circling and
emitting a sound similar to that of a suction pump, while feeding on the
abundant krill. It should be noted that the whales were sighted at the same
hours during the summer. The sightings were recorded in the morning at
08 : 00, in the afternoon at 14 : 30, in the evening at 20 : 30, and sometimes
also at 02 : 00 hs.

    Krill (*Euphausia sp.*) ———→ Fin Whale

V. CONCLUSIONS

On the basis of these observations, the sequence of the different groups of animals which are linked to the food chains in marine waters of Paradise Harbor is shown. When the conditions favorable to the boom of phytoplankton are present, the zooplankton begin to appear, initially by slow vertical migration from the surface to deep waters and vice versa. Many of these forms are confined to comparatively narrow layers in surface waters, which could be observed by diving in the area. Such distribution, adapted to definite conditions in temperature and salinity, is disturbed by the sea currents which make the organisms drift northward in surface waters. The zooplankton distribution influences that of other organisms which feed on it, such as all kind of invertebrates, fish, pinnipeds, cetaceans and birds.

Thus, it is most important to carry out a comprehensive investigation on the abiotic and biotic variations that affect the bioproduction which in turn influences the bioeconomics of the Antarctic Ocean. If further investigations are conducted in this way, the question of penguin and seal migrations in the Antarctic could be elucidated, since neither these animals nor the organisms on which they feed overpass the Antarctic Convergence.

So far, this problem was treated by the method of marking and recapture, which led to a minor success owing to errors in the procedure such as:

1) The animals are not marked on a large scale.

2) Along the continent and in the Subantarctic Islands there are not enough biological stations to capture or sight animals. Hence, the tags might be lost and if any marked animal is seen, the itinerary traced might be ignored. Thus, for instance, if a bird is ringed in Adelie Coast and later on is sighted in the South Orkney Island, it is difficult to determine whether it came from the east, the west, or crossed the Antarctic Continent. To conclude, I quote N. A. Mackintosh (1937) who demonstrated that the three predominant species of plankton in the Southern Ocean: *Rhincalanus gigas, Calanus acutus* and *Eicronia hamata* migrate to the north in surface waters in summer; then, they descend toward waters drifting southward and finally return to their primitive location. F. C. Fraser (1936) demonstrated that migration of *Euphausia sp.* in the larval form is congregated in deep temperate waters drifting southward. On reaching the ice edge the organisms rise to the surface following the vertical currents whereas the adults drift northward in surface waters.

These quotations corroborate the abovementioned regarding the importance of undertaking studies on the migration of birds, pinnipeds and cetaceans based on their main food distribution.

References

FRASER, F. C. (1936). On the development and distribution of the young stages of krill (*Euphausia superba*). *Discovery Repts.*, **14**, 3–192.

MACINTOSH, N. A. (1937). Seasonal circulation of the Antarctic macroplankton. *Discovery Repts.*, **16**, 367–412.

NEWELL, G. E. and NEWELL, R. G. (1963). *Marine Plankton—A Practical Guide.* Hutchinson Biological Monographs. Essex.

RAYMONT, J. E. G. (1963). *Plankton and Productivity in the Oceans.* International Series of Monographs on Pure and Applied Biology. Division Zoology. Vol. 18. New York.

Size distribution of the phytoplankton and its ecological significance in tropical waters

J. G. TUNDISI

Instituto Oceanográfico da Universidade de São Paulo

Abstract

In this paper, there are discussed some patterns of distribution of the phytoplankton in different size classes in tropical inshore and oceanic waters. The share of each fraction of the phytoplankton, the nannophytoplankton and the microphytoplankton was considered both as its contribution to the standing-stock and to the primary production of the total phytoplankton. A latitudinal comparison of the nanno/microphytoplankton contribution was also included. Some possible ecological implications at the primary and secondary trophic level are discussed.

Resumo

A distribuição do fitoplancton em diferentes classes de tamanho em águas oceânicas, costeiras e estuarinas em regiões tropicais é discutida. Deu-se ênfase às contribuições do nanofitoplancton e do microfitoplancton ao "standing-stock" e a produção de matéria orgânica total do fitoplancton.

Uma comparação da distribuição em tamanho do fitoplancton em diferentes latitudes é incluída. Possíveis implicações ecológicas, relacionadas com a rêde trófica, são discutidas.

---

* Portions of this paper were part of a thesis submitted in the University of S. Paulo, Brazil for the degree of Ph. D.

† Contribution No. 290 from the I.O. U.S.P.

INTRODUCTION AND GENERAL PROBLEMS

Studies on the size distribution of the phytoplankton in different areas of the Ocean can provide a useful background for a better knowledge of some interrelationships between the standingstock, the primary production, and certain environmental factors. Furthermore, these investigations have a considerable interest in the assessment of the quality and quantity of the food available to the herbivore zooplankton and suspension feeders in general. The purpose of the present paper is to review and discuss some of the available information on the size distribution of phytoplankton in tropical inshore and oceanic waters, and to compare these data with the investigations carried out in temperate waters. Some possible ecological implications of the patterns of size distribution will be discussed.

Methods

To fractionate the particulate material and phytoplankton, nylon or wire screens were generally used. Filters of various pore size have been used, but some problems arise with this material. During filtration, fragmentation of the more fragile cells can occur because of the effect of the vacuum (Holmes and Anderson, 1963; Anderson, 1965; Tundisi and Teixeira, 1968). Centrifugation has been used, to some extent to evaluate the smaller forms of the nannophytoplankton, but this seems to be a rather difficult method for routine analysis (Ballantine, 1953). In the last few years, the use of a Coulter Counter to estimate the size distribution has been preferred (Sheldon and Parsons, 1967; Parsons *et al.*, 1967, 1969). This method has the advantage of providing for automation of the technique, but one problem, is that it is not possible to separate living and non living material. In view of some of the problems discussed above, it seems that the estimation of the different size fractions of the phytoplankton in a given area, should be made by various techniques simultaneously. Microscopic examination of the phytoplankton and its fractions should also be included.

TROPICAL OCEANIC WATERS

Wood and Davis (1956), Teixeira (1963), Holmes and Anderson (1963), Saijo (1964), Saijo and Takesue (1965) have determined the contributions of the microphytoplankton and nannophytoplankton to the total chlorophyll content, cell numbers, and total C^{14} uptake. It was shown in these papers that the nannophytoplankton represents from 50–80 % of the total standing-stock and C^{14} uptake of the phytoplankton. Table I, taken from Teixeira (1963) shows the difference in C^{14} uptake of the nanno and the microphytoplankton at some stations in Equatorial Atlantic. At the stations located in the coastal water (stations 1, 2 and 4 in Table I) the microphytoplankton presented greater percentage as a component of the standing-stock of the total phytoplankton. When these data are compared with the C^{14} uptake of each fraction, it appears that the contribution of the nannophytoplankton is even greater than it would appear by its contribution to the standing-stock. At the offshore stations (3, 5, 6 in Table I), the nannophytoplankton dominated the samples as primary producer and as a part of the standing-stock.

Table 1 Phytoplankton standing-stock and C-14 uptake by net-phytoplankton and nannoplankton from surface sea-water samples collected in Equatorial waters. From Teixeira, (1963)

Sta. No.	Date	Latitude	Longitude	Relative photosynthesis		Relative standing-stock	
				Nanno-plankton	Netphyto-plankton	Nanno-plankton	Netphyto-plankton
1	22/3/63	04° 06.3 N	44° 44.0 W	87.00	13.00	71.50	28.50
2	22/3/63	03° 02.0 N	45° 21.0 W	88.30	11.70	73.70	26.30
3	22/3/63	01° 05.0 N	46° 23.8 W	90.60	9.40	77.90	22.10
4	22/3/63	00° 39.8 N	46° 38.0 W	89.00	11.00	78.00	22.00
5	22/3/63	00° 03.0 N	46° 59.0 W	91.60	8.40	82.70	17.30
6	22/3/63	47° 10.6 W	47° 10.6 W	93.60	6.10	83.00	17.00

* netphytoplankton = microphytoplankton.

Mullin (1965) determined the size distribution of particulate organic carbon in the waters Indian Ocean. It was found that the carbon content of the particle size from 1 μ to 10 μ, attained about 58% of the total carbon. It was suggested that most of the carbon in the samples was due to phytoplankton rather than to heterotrophs and detritus. McAllister, Parsons and Strickland (1960) have shown that most of the primary production of the phytoplankton in the Equatorial Pacific and 75% of the chlorophyll *a* in a water sample was due to the phytoplankton smaller than 10 μ.

The work carried out by Saijo (1964), Saijo and Takesue (1965) reported on the size distribution of photosynthesizing phytoplankton in the Indian

Fertility of the Sea

Table 2 Size distribution of photosynthesizing phytoplankton in the Indian Ocean (from Saijo, 1964)

Station	Depth (m)	Millipore filter			Net	Total
		HA 0.45~0.8 μ	AA 0.8~5 μ	SM 5~110 μ	XX 13 < 110 μ	
1 3° 22.5' N, 79° 07.6' E	0	15%	36%	46%	3%	100%
2 4° 58.9' S, 78° 03.4' E	0	37	36	27	0	100
	50	0	28	70	2	100
3 10° 55.5' S, 78° 07.0' E	0	25	26	49	0	100
	50	0	40	60	0	100
4 19° 54.1' S, 78° 01.0' E	0	5	41	48	6	100
	50	0	5	85	10	100
5 24° 58.7' S, 77° 59.6' E	0	33	3	52	12	100
	50	0	25	71	4	100
6 19° 50.2' S, 86° 11.9' E	0	2	35	63	0	100
	50	0	11	78	11	100
7 12° 44.3' S, 97° 20.0' E	0	16	47	37	0	100
	50	10	16	74	0	100
Mean value	0	19	30	48	3	100
	50	2	20	73	5	100

Ocean. It was found that the organisms with the size between 5 μ and 110 μ play the largest part of the activity. Also, the activity of organisms in larger fractions increased with depth. Table II taken from Saijo (1964) summarizes the results obtained at some stations. Another point considered was the relation between the size of the fraction, and the rate of photosynthesis by unit amount of chlorophyll. The highest value, was found in the fraction retained by the A. A. Millipore filter (i.e. the fraction greater than 0.8 μ). The lowest value, was found in the fraction greater than 110 μ.

TROPICAL INSHORE WATERS

Few investigations have been carried out in this environment. Most of the results available, are from a mangrove environment of the estuarine type at Cananéia (25° South Latitude) where estimates of the contribution of each fraction of the phytoplankton were studied during 1965 (Teixeira, Tundisi and Santoro, 1967). The data taken every other month at one station located in the inner parts of the mangrove area showed that the nanno-phytoplankton is responsible for most of the C^{14} uptake (Table III).

However, more detailed and intensive work (Tundisi, 1969, unpublished) showed that during the summer the percentage of the microphytoplankton in the total standing-stock increases as well as the C^{14} uptake of this fraction. During other times of the year the nannophytoplankton dominates the samples. The increase in microphytoplankton during the summer was attributed to better growth conditions due to nutrient accretion through rainfall and land drainage.

Simultaneous investigations of the zooplankton in this area showed that its bulk is composed of small copepods and larvae of benthic animals which would extensively use the nannophytoplankton. However, during the summer, part of the phytoplankton may escape predation due to its larger size.

Some results from the estuary of the Vellar River (India) (Savage, personal communication) confirm the predominance of the nannophytoplankton in this estuary. However, this investigation was limited to a short time in the year.

SIZE DISTRIBUTION OF PHYTOPLANKTON AT DIFFERENT LATITUDES

From the results given by the literature (Table IV), it appears that the microphytoplankton contributes a greater percentage of the standing-stock of the phytoplankton at higher latitudes. However, this latitudinal

Table 3 Nannophytoplankton and net-phytoplankton from surface and depth, in per cent, and hydrographical data. From Teixeira, Tundisi and Santore Ycaza, (1967)

Month	Nannophytoplankton (%)		Net-phytoplankton		Temperature °C		Salinity °/oo		Oxygen cc/L		pH		Precipitation high in mm/5d	Transparency in K=1.7/5	Tide
	Surface	Depth*	Surface	Depth*	Surface	Depth*	Surface	Depth*	Surface	Depth*	Surface	Depth*			
February	95.28	84.41	4.72	15.59	28.80	27.90	4.78	4.80	—	—	6.0	6.0	9.1	4.20	Low
April	95.83	90.56	4.17	9.44	24.80	24.50	15.00	14.65	2.91	3.00	6.0	6.0	9.0	2.70	Low
June	95.75	90.64	4.25	9.36	21.10	21.10	13.24	14.31	3.32	3.48	6.5	6.5	11.7	2.10	Low
August	82.32	64.32	17.68	35.68	22.10	22.10	19.92	19.84	4.23	4.18	6.5	6.5	0.4	1.30	High
October	94.34	80.00	5.66	20.00	23.42	23.38	22.61	23.22	4.37	4.40	6.5	6.5	12.7	1.40	Low
December	87.93	71.37	12.07	28.63	25.75	26.15	8.53	8.95	3.78	3.80	6.0	6.0	18.2	1.10	High

* 25% light penetration.
** Net-phytoplankton = microphytoplankton.

difference is probably valid for oceanic waters where little influence of such factors as a "land mass effect" occur. In coastal or inshore waters more eutrophic conditions at certain times of the year are responsible for a increase in the microphytoplankton due to optimal growth conditions.

Table 4 Standing-stock of the microphytoplankton as a percentage of the total phyto-plankton at different latitudes

Locality	Author	Method	Approximate Latitude	Microphyto-plankton
Scoresby Sound Groeland	Digby 1953	Membrane filter 0,6 μ, pore size	70° N	66%
English Channel	Harvey, 1950	Membrane filter	50° N	10–26%
Long Island Sound	Riley, 1941	Membrane filter	41° N	9–56%
Vineyard Sound (USA)	Yentsch and Ryther, 1959	Membrane filter	41° N	2–47%
Tortugas	Riley and Gorgy, 1948	Membrane filter	24° N	1%
Equatorial Atlantic	Teixeira, 1963	Millipore filter, H.A, 0.45 μ pore size	4° N	28%
Equatorial Atlantic	Teixeira, 1963	Millipore filter, H.A, 0.45 μ pore size	0°	17%
New South Wales	Wood and Davis 1956	Centrifugation	33° S	3–4%

SUMMARY AND DISCUSSION

From the results shown, and taking into account the problems of methodology involved, there seems to be reasonable evidence that the nannophytoplankton is the most important fraction, both as primary producer and as a component of the standing-stock of the phytoplankton in tropical waters.

According to Munk and Riley (1952) there are marked differences in the absorption times of nutrients by different cells of the phytoplankton, the small forms having the shortest absorption time. This is of fundamental importance in a environment of poor nutritional conditions as in tropical oceanic waters. Also, it must be taken into account that small cells capable of movement maintain themselves more easily in the euphotic zone in these waters. Some of the work carried out in tropical inshore and coastal waters showed a larger proportion of the standing-stock of the phytoplankton being constituted by the microphytoplankton during the summer. This might be due to nutrient accretion through run-off which is greater during this season. Also the introduction of humic substances in the system could be a factor of considerable importance in the growth of the microphytoplankton (Prakash and Hodgson, 1966; Teixeira, Tundisi and Santoro, 1969; Tundisi, 1969). The ecological implications of these differences in the size distribution of the phytoplankton, are mainly related to the spectrum distribution of the food available to the herbivore zooplankton and to the seasonal availability of nutrients throughout the year. As the food available is distributed mainly as particles of size smaller than 50 μ, the energy expediture of aggregating this food would be smaller for those species feeding on this size range. When the microphytoplankton contributes in greater percentage to the standing-stock of the total phytoplankton in inshore and coastal waters, there is the possibility that part of this fraction escape zooplankton predation, since small herbivores constitute this population. A short cut in the food chain in these tropical inshore waters, could thus be expected as demonstrated by Bainbridge (1963) and suggested by Smayda (1966), Quasim (1968), Teixeira, Tundisi and Santoro (1969).

It seems that the key to understanding the seasonal cycle and the factors regulating the growth of the different fractions of the phytoplankton are systematic experiments in the enrichment of the sea water throughout the year as well as experiments on the selective grazing of the zooplankton. This would give better indications on the basic aspects of the food chain at the primary and secondary trophic levels in tropical inshore and oceanic waters.

Acknowledgements

1. The author wishes to express his thanks to the "Fundação de Amparo à Pesquisa do Estado de S. Paulo" for providing funds.
2. Table I. Reprinted with permission from Clóvis Teixeira, Relative rates of photosynthesis and standing stock of the net phytoplankton and nannoplankton, 1963, Instituto Oceanográfico.
3. Table II. Reprinted with permission from Yatsuka Saijo, Size distribution of photosynthesizing phytoplankton in the Indian Ocean, 1964, Oceanographical Society of Japan.
4. Table III. Reprinted with permission from Clóvis Teixeira, Plankton studies in a mangrove environment IV. Size fractionation of the phytoplankton, 1967, Instituto Oceanográfico.

References

ANDERSON, G. C. (1965). Fractionation of phytoplankton communities off the Washington and Oregon coast. *Lim.* and *Oceanogr.* **10**, 477–479.

BAINBRIDGE, V. (1963). The food, feeding habits and distribution of the bonga Ethmalosa dorsalis (Cuvier & Valencienes). *J. Cons. Perm. Int. Explor. Mer,* **28**, 276–284.

BALLANTINE, D. (1953). Comparison of the different methods of estimating nannoplankton. *J. Mar. Biol. Ass. U. K.,* **32**, 129–142.

DIGBY, P. B. S. (1953). Plankton production in Scoresby Sound, East Greenland. *J. Anim. Ecol.,* **22**, 289–322.

HARVEY, H. W. (1950). On the production of living matter in the sea. *J. Mar. Biol. Assoc. U. K.,* **29**, 97–136.

HOLMES, R. W. and ANDERSON, G. C. (1963). Size fractionation of C^{14} labelled natural phytoplankton communities. In: *Symposium on marine microbiology.* Springfield, Thomas.

MULLIN, M. M. (1965). Size fractionation of particulate carbon in the surface waters of the Indian Ocean. *Lim.* and *Oceanogr.,* **10**, 459–461.

MCALLISTER, C. D., PARSONS, T. R. and STRICKLAND, J. D. H. (1960). Primary productivity at Station "P" in the North Coast Pacific Ocean. *J. Cons. Int. Expl. Mer,* **25**, 240–251.

MUNK, H. W. and RILEY, H. G. (1952). Absorption of nutrients by aquatic plants. *J. Mar. Res.,* **11**, 215–240.

PARSONS, T. R., LE BRASSEUR, R. J. and FULTON, D. J. (1967). Some observations on the dependence of zooplankton on the cell size and concentrations of phytoplankton blooms. *J. Oceanogr. Soc. Japan,* **23**, 10–17.

PARSONS, T. R., STEPHENS, K. and LE BRASSEUR, R. J. (1969). Production studies in the Strait of Georgia. Part I. Primary production under the Fraser River Plume, February to May, 1967. *J. Exp. Mar. Biol. Ecol.,* **3**, 20–39.

PRAKASH, A. and HODGSON, M. (1966). Physiological Ecology of marine dinoflagellates.— *Bedford Institute of Oceanography. Fifth Annual Report,* 93–94.

QASIM, S. Z. (1968). Some problems related to the food chain in a tropical estuary. *Symposium on marine food chains, University of Arrhaus, Denmark* (in press).

RILEY, G. A. (1941). Plankton Studies. III. Long Island Sound. *Bull. Bingham Oceanogr. Coll.,* **7**, 1–93.

RILEY, G. A. and GORGY, S. (1948). Quantitative studies of summer plankton populations of the Western North Atlantic. *J. Mar. Res.,* **7**, 100–121.

SAIJO, Y. (1964). Size distribution of photosynthesizing phytoplankton in the Indian Ocean. *J. Oceanogr. Soc. Japan,* **19**, 187–189.

SAIJO, Y. and TAKESUE, H. (1965). Further studies on the size distribution of photosynthesizing phytoplankton in te Indian Ocean. *J. Oceanogr. Soc. Japan,* **20**, 264–271.

SHELDON, R. W. and PARSONS, T. R. (1967b). A continuous size spectrum of the particulate matter in the sea. *J. Fish. Res. Bull. Can.,* **24**, 909–915.

SMAYDA, T. J. (1966). A quantitative analysis of the phytoplankton of the Gulf of Panama. III. *Inter American Tropical Tuna Commission,* **II**, 355–612.

TEIXEIRA, C. (1963). Relative rates of photosynthesis and standing-stock of the net-phytoplankton and nannoplankton. *Bolm. Inst. Oceanogr., Univ. S. Paulo,* **13**, 53–60.

20*

TEIXEIRA, C., TUNDISI, J. and SANTORO, Y. J. (1967). Plankton studies in a mangrove environmental. IV. Size fractionation of the phytoplankton. *Bolm. Inst. Oceanogr., Univ. S. Paulo*, **16**, 39–42.

TEIXEIRA, C., TUNDISI, J. and SANTORO, Y. J. (1969). Plankton studies in a mangrove environment. VI. Primary production, zooplankton standing-stock and some environment factors. *Int. Revue. Ges. Hydrob. Hydrogr.*, **54**.

TUNDISI, J. and TEIXEIRA, C. (1968). Plankton studies in a mangrove environment. VII. Size fractionation of the phytoplankton: Some studies on methods. *Bolm. Inst. Oceanogr., Univ. S. Paulo*, **17**, 89–94.

TUNDISI, J. (1969). Produção primária, "standing-stock" e fracionamento do fitoplancton na Região Lagunar de Cananéia. Ph. D. Thesis, University of S. Paulo (unpublished).

WOOD, E. F. J. and DAVIS, D. S. (1956). Importance of smaller phytoplankton elements. *Nature, Lond.*, **177**, 438.

YENTSCH, C. S. and RYTHER, J. H. (1959). Relative significance of the net-phytoplankton and nannoplankton in the waters of Vineyard Sound. *J. Cons. Perm. Int. Explor. Mer.* **14**, 231–238.

Quantitative studies on saprophytic phycomycetes in the South Atlantic near Santos (Brazil)

ANNEMARIE ULKEN*

Department of Botany, "Institut für Meeresforschung"
Bremerhaven, Germany

Abstract

During July in the Atlantic Ocean near Santos (Brazil) quantitative studies on the occurrence of saprophytic phycomycetes were made. In water samples the number of units of lower fungi was between 2 and 32 per l. In sediment samples the quantity varied between about 500 and 15,000 units per l. The number of fungi in both water and sediment decreases with the distance from the shore. Besides the number in sediment samples decreases with increasing depth.

Zusammenfassung

Im Juli wurden im atlantischen Ozean vor der brasilianischen Küste bei Santos quantitative Untersuchungen über das Vorkommen von saprophytischen Phycomyceten gemacht. In den Wasserproben schwankten die Zahlen zwischen 2 und 32 pro Liter. In den Sedimentproben befanden sich zwischen rund 500 und 15000 infektiösifen Einheiten pro Liter. Die Zahl ändert sich mit der Entfernung von der Küste und bei den Bodenproben mit der Tiefe. Nahe der Küste und in geringeren Tiefen sind die Zahlen höher als in der Hochsee und in größeren Tiefen.

---

* Dedicated to Prof. Dr. H. Engel for his 70th anniversary

The widely spread biflagellate fungi of the genera Thraustochytrium and Schizochytrium have been found in the North Atlantic (Sparrow 1934, 1968; Goldstein 1963; Gaertner 1966) and in the equatorial Atlantic (Ulken, 1966). These fungi are known to live on decaying algae (Sparrow, 1968) and detritus of higher marine plants (Goldstein, 1963) and thus play a role in mineralizing organic substances in the ocean. Gaertner (1966) inaugurated quantitative studies of these organisms in the North Atlantic and the North Sea, which is very rich in organic matter. As no such studies were conducted in the tropical South Atlantic the present quantitative investigation was undertaken.

MATERIAL AND METHODS

During a cruise of the Brazilian research vessel *Prof. W. Besnard* in July 1969 from Santos to the Ilha Grande water samples were collected with Nansen bottles, and sediment samples with a grab of the van-Veen type. The samples were treated immediately after collection by the method of Gaertner (1966). The baited samples were incubated in the dark at room temperature (about 20°C) for 14 days after which they were investigated microscopically for the first time. The culture flasks were then sent to Germany and observed for a second time 4 months later. At this time some of the samples which showed positive growth at the previous observation in Brazil gave negative results; on the other hand growth of Phycomycetes was found in some samples which seemed to be negative earlier. The results of all the positive samples are presented in Table 1.

RESULTS AND DISCUSSION

The surface water is very poor in lower fungi, as shown in Table 1. During the four months period of incubation in many of the cultures a growth of higher fungi (yeasts, fungi imperfecti) occurred. The spores of these fungi are probably of terrestrial origin. Sometimes these fungi suppressed the growth of the Phycomycetes. The number of Phycomycetes is high in the sediment and decreases as the depth increases. There are more fungi in samples near the shore than in those far offshore in the ocean. This correlates well with the results reported by Gaertner (1968). As the dilution method of Gaertner (1968) is very good for the isolation of single spores, some of these cultures are kept for taxonomical studies, which will be published later.

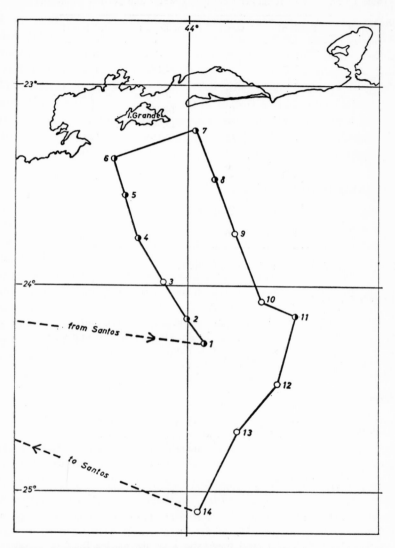

Figure 1 Cruise of the research vessel *Prof W. Besnard* in July 1969.
0 = water sample; ◑ = water and .sediment sample

Table 1 Number of fungi per l of surface water and per l of sediment sample

Station	water	sediment	depth (m)
1	32	1,030	220
2	2	2,195	150
3	2		
4	8	4,890	94
5	10	14,670	68
6	20	13,700	46
7	16	2,195	45
8	16	4,890	70
9	16		
10	24		
11	32	538	200
12	10		
13	8		
14	6		

The position of the stations is shown in fig. 1.

Acknowledgements

I would like to express my sincere thanks to Dr. M. Vannucci, Director of the Instituto Oceanografico do Universidade de São Paulo, Dr. L. Tommasi and all the colleagues of the Seção Quimica, who helped me with enthusiasm.

The research was supported by a travel grant of Deutsche Forschungsgemeinschaft.

References

GAERTNER, A. (1966): Vorkommen Physiologie und Verteilung "Mariner niederer Pilze" (Aquatic Phycomycetes). *Veröff. Inst. Meeresforsch. Bremerh. Sonderband* 2, 221–236.

GAERTNER, A. (1968a): Eine Methode des quantitativen Nachweises niederer, mit Pollen köderbarer Pilze im Meerwasser und im Sediment. *Veröff. Inst. Meeresforsch. Bremerh., Sonderband* 3, 75–89.

GAERTNER, A. (1968b): Die Fluktuationen mariner niederer Pilze in der Deutschen Bucht 1965 und 1966. *Veröff. Inst. Meeresforsch. Bremerh., Sonderband* 3, 105–117.

GOLDSTEIN, S. (1963): Studies of a new species of Thraustochytrium that displays light stimulated growth. *Mycologia* 55, 799–811.

SPARROW, F. K. (1936): Biological observations on the marine fungi of Woods Hole waters. *Biol. Bull.* 70, 236–263.

SPARROW, F. K. (1968). On the Occurrence of the Thraustochytriaceae. *Veröff. Inst. Meeresforsch. Bremerh.* 11, 89–91.

ULKEN, A. (1966). Sôbre a ocorréncia de fungos em amostras do Atlântico Equatorial. Instituto de Pesquisas da Marinha, *Nota Tecnica* 28, 1–11.

Production and prediction

E. J. FERGUSON WOOD

Institute of Marine Sciences
Miami, Florida

Abstract

Quantitative and qualitative phytoplankton data collected over approximately 30 years, together with synoptic hydrology and nutrient information are being analysed, first by multiple regression and second by recurrent group analysis. The area covered by this study includes the Antarctic, eastern Indian and South Pacific Oceans from 90–180° E, the tropical Atlantic, especially the western Atlantic and the Caribbean Sea.

Unfortunately, the analyses are incomplete at this time, but the kinds of information made available are, for example, the phosphate and nitrate are seldom if ever limiting in the Antarctic to phytoplankton growth, and that phytoplankton communities can be identified with water masses even when these sink to upwards of 1000 meters.

It is suggested that analyses of the kind being used in this study may be able to provide field models allowing of the prediction of qualitative and perhaps semi-quantitative blooms of phytoplankton where sufficient data are available. They might also serve as a clue for experimental studies and the matching of experimental and field models.

Resumen

Dados qualitativos e quantitativos de fitoplâncton coletados durante aproximadamente 30 anos, juntamente com dados sinóticos de hidrologia e nutrientes estão sendo analisados, primeiramente por regressão múltipla e segundo por análises recorrentes de grupos. A área coberta por êsse estudo, inclui o Antártico, os Oceanos Pacífico e Índico, Leste, de 90–180° E, o Atlântico Tropical, especialmente o Atlântico Oeste e o Mar das Caraibas.

Infelizmente, as análises ainda não estão completas, mas as informações mais evidentes são de que o fosfato e nitrato são raramente limitantes ao crescimento do fitoplâncton no Antártico e que comunidades do fitoplâncton podem ser identificadas com massas de água, mesmo quando elas afundam abaixo de 1,000 metros.

Prevê-se que análises como as que estão sendo feitas nesse estudo, possam servir para a construção de modelos de predição de crescimento explosivo de fitoplâncton, onde haja dados suficientes. Podem também servir como indicação para estudos experimentais e na integração entre modelos experimentais e de campo.

When one considers the large numbers of expeditions which have collected data on phytoplankton, zooplankton, hydrology and other relevant information, one would expect that there would be a large body of available information about the distribution of phytoplankton species, the composition of phytoplankton communities and the relationships of these to the natural habitat. In point of fact, very little information is available, and the raw data are still stored away in various laboratories and have been interpreted to a very limited extent.

On the other hand, especially in recent years, a large accumulation of data has been amassed on marine standing crop and "productivity", and a number of assessments of annual primary production has been made of the world oceans and of particular areas. It need hardly be said here that these assessments often differ by one or more orders of magnitude. For example, the Antarctic is often stated to be highly productive in the light of figures supplied by Burkholder and Mandelli (1965), El Sayed (1966) and others, while the Sargasso Sea is almost universally held to be a marine desert. However, Walsh (1960, 1969a) using Burkholder's and El Sayed's data concluded that there was little difference in annual productivity between the Antarctic and the Sargasso Sea, and Pomeroy and Weibe (1969), using different methods, came to the same conclusions. Further, considerable criticism has been leveled at the various methods of measuring either productivity or standing crop, and the errors of estimation may well amount to an order of magnitude.

In addition, there is a large sampling error which is generally believed to be greatly increased by patchiness. When I have pointed this out to my students, with due recognition of my authorities, they say quite correctly that this is purely destructive criticism, and ask whether I have no alternative to offer. In this paper, I am trying to offer, not exactly an alternative, but something which may prove to be a useful tool, and to lead to a better assessment of phytoplankton populations.

Firstly, let me state than I am unconvinced of the universality of patchiness as a major factor in oceanic environments. Most of the work done on this has been done in estuarine situations such as that of Cassie (1962) which was done primarily in Hauraki Gulf where even a human may find himself in water of several rather different temperatures at the same time, in other words the patchiness is in the water environment rather than in the phytoplankton. I have taken samples at half hour intervals for some 300 miles and at hourly intervals from an anchored ship at the edge of a continental shelf in a 3-knot current (the East Australian current) and the variation in numbers was not greater than that which one could expect from diurnal variation. This applied to the total number of organisms estimated by fluorescence microscopy through the photic zone. I found also that the coefficient of variation of duplicate samples examined by this method was of the order of 20–30 %, usually around 28 %. However, in the case of surface red tides such as those of *Trichodesmium* which normally are confined to the first 50 cm, patchiness tends to be greater, as these are affected by surface phenomena including tides, wave action and winds, and a circulation, probably of the Langmuir variety is bound to occur.

Anderson (1969) reported that there is in a large part of the northern Pacific a chlorophyll maximum in the vicinity of 60 meters depth. I had recorded this (Wood, 1964) regarding phytoplankton numbers from the Indian, Pacific and Antarctic waters, and have confirmed it for the tropical Atlantic, Mediterranean Sea and Caribbean Sea. There are papers which confine discussion of phytoplankton productivity to the first 50 meters, based on compensation point data, and these are, of course, of little value.

While I do not regard any of the present methods of assessing standing crop or assimilation as being highly accurate, (in fact they give merely a relative picture), I believe that they are very useful for giving a general ecological picture of oceans and seas, and that they can be very informative.

Mathematical models for productivity systems have, in the past, been largely independent of field evidence, and for this reason have been applicable only in the cases for which they were invented. My group has been trying to use the very large amount of phytoplankton and associated data which I have collected over the years in order to produce a model from field evidence and to adapt this to make a realistic appraisal of phytoplankton distributions in the world oceans, not so much from the quantitative point of view as the qualitative or semiquantitative; i.e. to predict when and where phytoplankton blooms are likely to occur and under what conditions, as well as the species or community likely to be involved.

My student, Dr. John J. Walsh (1969b) has made a preliminary study of our Antarctic data, and we had planned, for this symposium, to add to this, analysed data from the eastern Indian Ocean from the equator to the southern edge of the subtropical convergence and from 90° E to the west coast of Tasmania and Torres Strait. However, computer troubles have delayed our schedule, so much of the information is not yet available. We have, however, found some very interesting things; in the Antarctic, by a study of multiple regression analyses, we have come to the conclusion that;

a Having data on chlorophyll, carbon 14 assimilation and phytoplankton counts does not add more information to the equations than any one of these alone. This suggests, as I have often thought from purely biological considerations, that ^{14}C studies yield information on standing crop rather than on dynamic productivity.

b Phosphate and nitrogen are not limiting in the Antarctic, except perhaps on very rare occasions.

c Light is not limiting until it is less than 10% (probably much less) of surface illumination.

d Both quantitatively and qualitatively, phytoplankton is closely related to sigma T curves, and thus to water masses. Thus a given phytoplankton population can be recognised in a sinking water mass to a depth of about 1200 meters and probably more, though the number of organisms, and also their diversity, will decrease greatly.

We now propose to use the multiple regression analyses to try to interpret the distribution of phytoplankton in the south-west Pacific and tropical Atlantic Oceans and the Caribbean.

In addition to this, we are not developing a concurrent group analysis program in order to study community structure as related to oceanic habitat. We hope, by this method to gain information on the relation of pecies groups to habitat, and some indications of factors influencing succession. We shall also find out the imperfections of our data, and may be an indication of where data can be improved.

One difficulty is that, since we are unable to grow many of the important oceanic species in culture, especially in defined media, we cannot readily assess the factors which are likely to be missing from our equations. A sad lack, and one which we fully recognize, is what may be called the debit side of our ledger. This included grazing, sinking (if appreciable), and death. I have only eight 24 hour stations from which diurnal grazing and replenishment of crop might be assessed, and we know that a large number of such stations is required in each area of study. This type of station entails staying for over 24 hours in the same water mass and sampling at intervals of not more than 4 hours, which is not easy to do in practice especially as most ship cruises are multi-purpose. It is not necessary in the first instance to separate the three factors mentioned, as they are all working towards the same end-reduction of the population. Later on of course, separation is desirable.

We believe that the methods we are using for studing our own data and that of other expeditions such as the *Discovery* and *William Scoresby* will allow us to make some assessment of the apparent causes of phytoplankton growth and limitation, and to suggest changes in the type of information required to improve our data.

I am particularly interested in the region to the west of the Amazon River, and the water extending from there into the Caribbean Sea, and Straits of Florida and ultimately forming the main part of the Gulf Stream. I have studied the part played by the fresh water of the Amazon in the phytoplankton distribution between there and the Windward Islands, though we have not yet had time to apply our statistical methods to this in any detail. It would appear that the Amazon has a suspisingly small effect on the phytoplankton distribution as a whole, except in a small area from the Amazon canyon to Surinam, and that the trade winds and the strong westerly movement of the Tropical South Atlantic water and the North Equatorial current have a very strong and persistent effect which controls the whole system, in the photic zone (Corcoran and Wood, M.S.).

To the east of the Windward Islands, upwelling plays a very large part in the distribution of phytoplankton, and this influence extends into the eastern Caribbean Sea. In a paper at the Curacao Meeting on the Caribbean, in 1968 (Wood, 1968), I gave a preliminary discussion of this, and to date, I have not had an opportunity to reassess my Caribbean data by the methods I have been discussing. I am convinced, however, that such a study will be of great value, especially as some channels between the West Indies islands are deep enough to cause little or no upwelling and to allow through the surface water (North Equatorial), subtropical underwater (which I now believe to be derived from the tropical South Atlantic water), the oxygen minimum layer (probably of Mediterranean origin) and part at least of the sub-Antarctic Intermediate water.

Lamont-Doherty Geological Observatory has a plan for creating an artificial upwelling in the Anagarda Passage on the north coast of St. Croix. If this, or a similar project could be used to create an artificial up-welling for large-scale experiments, it might be possible to use known parameters from field observations around the nearby islands to build a controllable model to simulate natural upwelling in the Caribbean region. The ultimate aim of such a research plan would be to predict qualitatively, and perhaps semi-quantitatively, the distribution of phytoplankton popu-lations and species assemblages there, and perhaps, *mutatis mutandis* in other areas, in which there are already a considerable amount of unstudied or partially studied data.

References

ANDERSON, G. C. (1969). On the distribution and development of a subsurface chlorophyll maximum in the northeast Pacific. Paper presented at the A.S.L.O. Meeting, La Jolla, 1969.

BURKHOLDER, P. R. and E. F. MANDELLI (1965). Carbon assimilation of marine phyto-plankton in Antarctica. *Proc. Nat. Acad. Sci.* **54**, 437–444.

CASSIE, R. M. (1962). Microdistribution and other error components of C14 primary production studies. *Limnol. Oceanogr.* **7**, 121–130.

EL SAYED, S. (1966). Phytoplankton production in Antarctic and Subantarctic waters. *2nd. Intern. Oceanogr. Symp.* Moscow.

POMEROY, L. R. and WIEBE, W. (1969). Variations in plankton respiration with depth and latitude. Paper presented at the A.S.L.O. Meeting at La Jolla, Sept. 1969.

WALSH, J. J. (1968). The vertical distribution of phytoplankton in the waters of the Palmer Peninsula. M. S. Thesis, Univ. of Miami, Coral Gables, Florida.

WALSH, J. J. (1969a). Vertical distribution of Antarctic phytoplankton, II. *Limnol. Oceanogr.* **14**, 86–94.

WALSH, J. J. (1969b). A statistical analysis of the phytoplankton community within the Antarctic convergence. Ph. D. dissertation, University of Miami, Coral Gables, Florida.

WOOD, E. J. F. (1964). Studies in microbial ecology of the Australasian region. *Nova Hedwigia*, **8**, 1–54; 453–568.

WOOD, E. J. F. (1968). Phytoplankton distribution in the Caribbean Region. Paper presented at C.I.R.C.A.R. Meeting. Curacao, 1968.